Research and Perspectives in Neurosciences

For further volumes:
http://www.Springer.com/series/2357

Bart de Strooper • Yves Christen
Editors

Macro Roles for MicroRNAs in the Life and Death of Neurons

Springer

Editors
Dr. Bart de Strooper
KULeuven and VIB
Center Human Genetics
Lab. Neuronal Cell Biology
Herestraat 49
3000 Leuven
Campus Gasthuisberg 6
Belgium
Bart.destrooper@cme.vib-kuleuven.be

Dr. Yves Christen
Fondation IPSEN pour la
Recherche Therapeutique
65 quai George Gorse
92650 Boulogne Billancourt
Cedex
France
yves.christen@beaufour-ipsen.com

ISSN: 0945-6082
ISBN: 978-3-642-26204-3 e-ISBN: 978-3-642-04298-0
DOI 10.1007/978-3-642-04298-0
Springer Heidelberg Dordrecht London New York

Cover design: WMXDesign GmbH, Heidelberg, Germany

Printed on acid-free paper

Springer is part of Springer Science+Business Media (www.springer.com)

Foreword

The discovery of microRNAs has revealed an unexpected and spectacular additional level of fine tuning of the genome and how genes are used again and again in different combinations to generate the complexity that underlies for instance the brain. Since the initial studies performed in C.elegans, we gave gone a far way to begin to understand how microRNA pathways can have an impact on health and disease in human. Although microRNAs are abundantly expressed in the brain, relatively little is known about the multiple functions of these RNA molecules in the nervous system. Nevertheless, we know already that microRNA pathways play major roles in the proliferation, differentiation, function and maintenance of neuronal cells. Several intriguing studies have linked microRNAs as major regulators of the neuronal phenotype, and have implicated specific microRNAs in the regulation of synapse formation and plasticity. Dysfunction of microRNA pathways is also slowly emerging as a potential important contributor to the pathogenesis of major neurodegenerative disorders such as Alzheimer's disease and Parkinson's disease. These novel insights appear to be particular promising for the understanding of the very frequent and badly understood sporadic forms of these diseases as compared to the genetic forms. Thus, the better understanding of the implications of this novel field of molecular biology is crucial for the broad area of neurosciences, from the fundamental aspects to the clinic, and from novel diagnostic to potentially therapeutic applications for severe neurological and maybe psychiatric diseases. The present volume gathers contributions to the Colloque Médecine et Recherche on the implications of microRNAs in neuroscience organized by the Fondation Ipsen, in Paris, on April 20, 2009. It had as objective to bring together neuroscientists from different areas of research to discuss their current insights into the wonderful world of microRNAs, and to hear and discuss their research and views about microRNA biology in neuronal processes and in brain disorders.

Bart de Strooper
Yves Christen

Acknowledgments

The editors wish to thank Jacqueline Mervaillie and Sonia Le Cornec for the organization of the meeting and Mary Lynn Gage for the editing of the book.

Contents

Contributors

Abeliovich Asa Columbia University Medical Center, 630 West 168th Street, Room 15-405, New York, NY 10032, USA, aa900@columbia.edu

Abu-Elneel Kawther Neuroscience Research Institute, Department of Molecular Cellular and Developmental Biology, University of California Santa Barbara, USA

Bonini Nancy M. University of Pennsylvania, 306 Leidy Laboratories, Department of Biology, Philadelphia, PA 19104, USA, nbonini@sas.upenn.edu

Chang Shuang Department of Human Genetics, Emory University School of Medicine, Atlanta, GA 30322, USA

Cohen Stephen M. Temasek Life Sciences Laboratory Limited, 1 Research Link National University of Singapore, 117604 Singapore, SINGAPORE, steve@tll.org.sg

De Strooper Bart Center for human genetics, K.U. Leuven and Department of molecular and developmental genetics, VIB Leuven, BELGIUM

Gogos Joseph A. Department of Physiology & Cellular Biophysics and Department of Neuroscience, Columbia University, New York, USA

Hébert Sébastien S. Centre de Recherche du CHUQ (CHUL), Axe Neurosciences, Université Laval, Département de Biologie médicale, 2705 Boul. Laurier, Local RC-9800, Québec, Qc, Canada, sebastien.hebert@crchul.ulaval.ca

Jin Peng Department of Human Genetics and Graduate Program in Genetics and Molecular Biology, Emory University School of Medicine, Atlanta, GA 30322, USA, pjin@genetics.emory.edu

Kapsimali Marika INSERM U784, Génétique Moléculaire du Développement, Ecole Normale Supérieure, 46 rue d'Ulm, 75230 PARIS Cedex 05 FRANCE, kapsimal@biologie.ens.fr

Karayiorgou Maria Columbia University, Department of Psychiatry, 1051 Riverside Drive, Unit #28, New York NY 10032, USA, mk2758@columbia.edu

Kosik Kenneth S. Neuroscience Research Institute, Department of Molecular Cellular and Developmental Biology, University of California Santa Barbara, Biology II, Room 6139A, Santa Barbara, CA 93106, USA, kosik@lifesci.ucsb.edu

Kye Min Jeong Neuroscience Research Institute, Department of Molecular Cellular and Developmental Biology, University of California Santa Barbara, USA

Liu Tsunglin Neuroscience Research Institute, Department of Molecular Cellular and Developmental Biology, University of California Santa Barbara, USA

Mandemakers Wim Center for human genetics, K.U. Leuven and Department of molecular and developmental genetics, VIB Leuven, BELGIUM

Papadopoulou Aikaterini S. Center for human genetics, K.U. Leuven and Department of molecular and developmental genetics, VIB Leuven, BELGIUM

Papagiannakopoulos Thales Neuroscience Research Institute, Department of Molecular Cellular and Developmental Biology, University of California Santa Barbara, USA

Rage Florence Institut de Génétique moléculaire de Montpellier, UMR 5535, 1919 route de Mende 34293 Montpellier Cedex 05, FRANCE, florence.rage@igmm.cnrs.fr

Schratt Gerhard M. University of Heidelberg, Interdisciplinary Center for Neurosciences (IZN), SFB488 Junior Group, Im Neuenheimer Feld 345, 1.OG, 69120 Heidelberg, GERMANY, schratt@ana.uni-heidelberg.de

Simon David J. Department of Molecular Biology, Massachusetts General Hospital, 185 Cambridge Street, Simches 7, Boston, MA 02114, USA, david.jason.simon@gmail.com

Szulwach Keith Department of Human Genetics and Graduate Program in Genetics and Molecular Biology, Emory University School of Medicine, Atlanta, GA 30322, USA

Xu Bin Department of Physiology & Cellular Biophysics and Department of Psychiatry Columbia University, New York, USA

Na Xu Neuroscience Research Institute, Departement of Molecular Cellular and Developmental Biology, University of California Santa Barbara, USA

Profiling the microRNAs

Kenneth S. Kosik, Thales Papagiannakopoulos, Na Xu, Kawther Abu-Elneel, Tsunglin Liu, and Min Jeong Kye

Abstract Profiling has contributed to the rapid growth of knowledge concerning microRNAs (miRNAs). The ability to measure all the known miRNAs by hybridization or by quantitative PCR has yielded insights into oncogenesis, stem differentiation, development, and disease. The steps toward discovery have often begun with profiling, followed by target prediction for miRNAs that are differentially expressed and then experimental validation of the targets. Here we present some methodological detail on the profiling procedures used in the lab and their application.

1 Introduction

microRNAs have emerged as a critical layer of post-transcriptional regulation over the expression of many genes (Bartel 2004; Stefani and Slack 2008). Among the super-kingdoms of life - Archaea, Bacteria, and Eukarya - microRNAs are found in Eukarya. Within Eukarya, they are present in the animal and plant kingdoms and absent in Fungi. Information about microRNAs remains scant in Protista.

microRNAs (miRNAs) regulate both the stability and translatability of their mRNA targets by binding to an imperfectly matched sequence within the target mRNA. Descriptions of miRNA function often focus on their role as post-transcriptional regulators of target mRNAs or on their organismal roles. However, a poorly understood territory lies between these biological levels of action. The effects of miRNA targeting are reflected in the proteome and, in some cases, ramify back through the transcriptional profile. Thus, mRNAs that may not be direct targets of miRNAs may exhibit more dramatic changes than their direct targets

K.S. Kosik (✉)
Neuroscience Research Institute, Departement. of Molecular Cellular and Developmental Biology
University of California Santa Barbara, Biology II Room 6139A, Santa Barbara, CA93106, USA

B. de Strooper and Y. Christen (eds.), *Macro Roles for MicroRNAs in the Life and Death of Neurons*, Research and Perspectives in Neurosciences,
DOI 10.1007/978-3-642-04298-0_1, © Springer-Verlag Berlin Heidelberg 2010

and may lie closer to the phenotypic consequences of miRNA inhibitory effects than direct targets.

miRNAs are particularly prevalent within the nervous system, where they contribute to the complexity and function of the mammalian brain (Mercer et al. 2008). One role of miRNAs in the nervous system appears to be related to plasticity and, more speculatively, to the epigenetic and transcriptional programs that underlie long-term memory storage.

To understand and develop full descriptions of these many facets of miRNA biology, detailed and sensitive profiling of miRNA levels is necessary, as well as mRNA profiles and the analytical tools to link these data sets.

2 The first miRNA profiling

We performed some of the earliest miRNA profiling by spotting membranes with tri-mer oligonucleotides (antisense to microRNAs) of 54–72 nt at a final concentration of 7 μM (Krichevskyet al. 2003). The oligonucleotides were spotted on the GeneScreen Plus (NEN) membranes with a 1536 pin plate replicator (V&P Scientific). To confirm the specificity of hybridization, a series of oligonucleotides with three mismatches (G → C or C → A) were included on the array. These mismatches resulted in a significant drop in signal from these spots compared to their cognates. Five to ten micrograms of total RNA from brain tissue was filtered through Microcon YM-100 concentrators to obtain a low molecular weight fraction of RNA enriched in molecules under 60 nucleotides. Despite this filtering, several highly expressed pre-miRNAs in the 70 nucleotide range were detected. A synthetic 21-nt RNA with a sequence that does not correspond to any miRNA, but is an exact complement to a random spotted sequence, was added to the RNA sample at a known concentration as a reference for normalization. The Low Molecular Weight (LMW) RNA was then end-labeled with 30 μCi of γ^{33}P dATP (3,000 Ci/mmole) by T4 polynucleotide kinase and purified. A typical experiment included three independent RNA samples for each experimental condition.

Among the findings we reported using this method was the highly consistent increase in levels of miR-21 in glioblastomas (Chan et al. 2005). Subsequently, miR-21 has been shown to be the microRNA most likely to be elevated in a wide range of solid tumors. In a follow-up study, we showed that miR-21 targets multiple important components of the p53, transforming growth factor-beta (TGF-beta), and mitochondrial apoptosis tumor-suppressive pathways (Papagiannakopoulos et al. 2008). Down-regulation of miR-21 in glioblastoma cells led to de-repression of these pathways, causing repression of growth, increased apoptosis, and cell cycle arrest. These findings clearly established miR-21 as an oncogene that targets a network of p53, TGF-beta, and mitochondrial apoptosis tumor suppressor genes in glioblastoma cells.

With this technique and additional supportive data, we also reported that brain-specific miR-124a and miR-9 molecules affect neural lineage differentiation in embryonic stem cell-derived cultures (Krichevsky et al. 2006) and published the first report of systematic miRNA gene profiling in cells of the hematopoietic system (Monticelli et al. 2005).

3 miRNA profiling by multiplex real time PCR

Commercialization of hybridization arrays with fluorescent dyes and glass slides came quickly. However, our profiling departed from the hybridization approaches and, in collaboration with Applied Biosystems, we profiled miRNAs using multiplex real-time RT-PCR. In one of our earliest experiments with this method, we used a 187-plex RT-PCR adapted for the rat sequences from Lao et al. (2006) to profile miRNAs in laser-captured cell bodies and neurites of dissociated hippocampal neurons (Kye et al. 2007). Briefly, multiplex reverse transcription primers were generated with a common 20 nucleotide stem loop followed by eight nucleotides complimentary to the 3′-end of a given miRNA (e.g., 5′-AACTATAC-3′for rno-let-7a). Reverse transcriptase reactions were performed in 5 μl containing 1x cDNA Archiving Kit buffer (Applied Biosystems), 10 units MMLV reverse transcriptase, 1.25 mM of each dNTP, 1.3 units RNase inhibitors (Applied Biosystems), and 2.5 nM of each reverse primer. Our sample RNA in this case was rat hippocampal dissociated culture taken by laser capture microdissection. Reactions were 30 cycles at 20°C for 30 sec, 42°C for 30 sec, and 50°C for 1 sec and one cycle at 85°C for 5 min to inactivate the enzyme.

cDNA was pre-amplified with a common reverse primer (UR) to the 20 nucleotide stem loop and multiplexed ∼18 nucleotide forward primers with a common 5′-end and a 3′-end matching 12-17 nucleotides of the 5′-end of a given miRNA (e.g., 5′-TCCAGCTCCTATATGAT-3′ for rno-let-7a). The amplification was performed in 25 μl containing 1x Universal Master Mix (Applied Biosystems), 50 nM of each forward primer, 5 μM of UR, 6.25 units AmpliTaq Gold polymerase (Applied Biosystems), 0.5 mM of each dNTP, 2 mM MgCl2, and 5 μl of the initial reverse transcriptase amplification. Reactions were one cycle at 95°C for 10 min, 55°C for 2 min, and 14 cycles at 95°C for 1 sec and 65°C for 1 min.

The product of this amplification was diluted four times and 0.1 μl was used for real-time PCR performed in 10 μl containing 1x Universal Master Mix (Applied Biosystems), 0.25 μM FP, 0.1 μM TaqMan® probe, and 5 μM UR. The miRNA specificity for each reaction was provided by the miRNA-specific sequences of the forward primers and TaqMan® probes. Real-time PCR was performed in a Biosystems 7500HT 96-well plate Sequence Detection System using 40 cycles of 95°C for 15 sec and 60°C for 1 min. All reactions were run in duplicate. The threshold cycle (Ct) was determined as the fractional cycle number at which the fluorescence passes the fixed threshold. All Ct values were averaged over two to four repeated biological experiments, each with two averaged real-time PCR experimental replicates. Most recently, Applied Biosystems decided to commercialize this product as a series of single-plex real-time PCR reactions and current methodologies differ slightly from what is described above.

A convenient facet of the multiplex approach is the flexibility to include mRNA detection in the same sample for real-time PCR. In our case, we used the same laser capture microdissected samples. Reverse transcriptase reactions were performed in 5 μl containing 1x cDNA Archiving Kit buffer (Applied Biosystems), 10 units MMLV reverse transcriptase, 1.25 mM of each dNTP, 1 unit RNase inhibitor (Applied Biosystems), 2 mM random primer, and 1 μl of the laser capture microdissection samples. Reactions were at 20°C for 10 min, 37°C for 120 min, and 85°C for 5 min to inactivate the enzyme. cDNA was pre-amplified using gene-specific forward primers and reverse primers. The amplification was performed in 25 μl containing 1x Universal Master Mix (Applied Biosystems), 50 nM of each forward primer, 50 nM of each reverse primer, 6.25 units AmpliTaq Gold polymerase (Applied Biosystems), 0.5 mM of each dNTP, 0.75 mM MgCl2, and 5 μl of the initial reverse transcriptase amplification. Reactions were one cycle at 95°C for 10 min, 55°C for 2 min, and 14 cycles at 95°C for 1 sec and 65°C for 1 min. The product of this amplification was diluted four times and 0.1 μl was used for real-time PCR reactions detecting mRNAs. It was performed in 10 μl containing 1x Universal Master Mix (Applied Biosystems), 500 nM forward primer, 200 nM TaqMan® probe, and 500 nM UR.

As noted above, the results from the miRNA profiling by quantitative PCR are given in Ct values. The cycle number at the threshold level of log-based fluorescence is defined as Ct number. Because larger Ct values tend to have greater fluctuation, we set 28 as the cutoff value above which the fluctuation makes precise determination of the Ct less reliable and below which the Ct value is highly reliable. This feature of the data can be clearly shown on a scatter plot in which two identical samples are profiled and the points where the line splays are consistently at a Ct value of 28. In a study of rat hippocampal neurons (Kye et al. 2007), among the 187 miRNAs in the raw data, 30 had an NTC Ct <28 (high NTC noise): 36 Cts were above the cut-off in all samples, four had a large fluctuation in duplicated experiments, and 13 had a large fluctuation between samples (s.d.>1). A further five miRNAs were excluded with a mean Ct in the cell body >28 even though their among-sample fluctuation was small. After filtering, we were left with 99 miRNAs.

Real-time reverse transcription PCR was useful for detecting a microRNA signal in an in vivo expression set of mRNAs (Liu et al. 2007). In this study, RNA was extracted from 12 human primary brain tumor biopsies. We determined genome-wide mRNA expression levels by microarray analysis and a miRNA profile by real-time reverse transcription PCR. Correlation coefficients were determined for all possible mRNA-miRNA pairs, and the distribution of these correlations was compared to the random distribution. Within the correlations we sought endogenous fluctuations in the data set that were statistically distinguishable from the many other fluctuations in the data set. In fact, an excess of high positive and negative correlation pairs was observed at the tails of these distributions. Among the 70 most highly correlated pairs (>0.9), we did not observe direct targets. This result suggested that, in a total transcriptional profile, the dominant effect of changes in the endogenous expression of miRNAs occurs on non-target mRNAs. We concluded that sufficient information exists within a set of tumor samples to detect endogenous correlations between miRNA and mRNA levels. From these data sets, we inferred and validated a tumor suppression pathway linked to miR-181c. We suspect that the detection of statistically significant correlations and anti-correlations in this large data set of 12 glioblastomas would not have been possible without the dynamic range of PCR-based detection of miRNA levels as opposed to the less sensitive array-based methods.

Recently, profiling of miRNAs has allowed us to discover that miR-145 has a repressive role on human embryonic stem cell pluripotency (Xu et al. 2009). In this study, profiling revealed that miR-145 was prominent among the up-regulated miRNAs when human embryonic stem cells transitioned to embryoid bodies. Based initially on target predictions and then on validated experimental data, we found that miR-145 targeted the "reprogramming factors" OCT4, SOX2 and KLF4. Increased miR-145 inhibited self-renewal, repressed expression of pluripotency genes, and induced lineage-restricted differentiation in human embryonic stem cells. Conversely, loss of miR-145 impaired stem cell differentiation and elevated OCT4, SOX2 and KLF4. Interestingly, the miR-145 promoter is bound to and repressed by OCT4 in human embryonic stem cells. Thus, a direct link exists between the core reprogramming factors of stem cells and at least one microRNA. This link operates as a double-negative feedback loop involving OCT4, SOX2, KLF4 and miR-145.

The multiplex real time PCR was also sensitive enough to detect some variation among a set of brain samples from individuals affected by autism spectrum disorder (Abu-Elneel et al. 2008). In this setting, the application of miRNA profiling is most challenging because miRNA changes are subtle. Individual differences in terminally differentiated tissues are few and minimal in magnitude. In contrast, the cells or tissues that are undergoing changes in their identity related to development, stem cell differentiation, or oncogenesis are readily detectable. Also challenging in complex genetic diseases such as autism is the diversity of genetic variation that can lead to the autism phenotype. Given these limitations, we compared the expression of 466 human miRNAs from postmortem cerebellar cortex tissue of 13 individuals with autism spectrum disorder and 13 controls from non-autistic cerebellar samples. Most miRNA levels showed little variation across all samples suggesting that autism does not induce global dysfunction of miRNA expression. However, some miRNAs among the autistic samples were expressed at significantly different levels compared to the mean control value. Although no set of miRNAs could be implicated as being across-the-board dysregulated among all cases, some miRNAs differed from controls in individual cases. This finding pointed to the well-recognized heterogeneity of autism. Interestingly, among the predicted targets of dysregulated miRNAs were genes that are known genetic causes of autism, such Neurexin and SHANK3.

4 In situ hybridization for miRNAs

In situ hybridization for miRNAs is fraught with pitfalls due to their short lengths and small differences in sequence, especially among closely related miRNAs. However, it is a useful validation tool. In our studies, in situ hybridization was used to confirm the localization of certain miRNAs to neuronal dendrites (Kye et al. 2007). Specifically, locked nucleic acid (LNA; Exiqon) probes were used to confirm the differential spatial distribution of rno-miR-26a and miR-124a.

LNA probes were labeled with biotin using the 3′-end DNA Labeling Kit (Pierce) and purified using chloroform/isoamylalcohol. The LNA probes had the following sequences: tggcattcaccgcgtgccttaa (rno-mir-124a), gcctatcctggattacttgaa (rno-mir-26a), and catgtcatgtgtcacatctctt as a negative control. This control does not match any sequence in the rat genome. Two LNAs were usually placed at each end of the sequence. Cultures were washed in PBS, fixed in 4% PFA in PBS/5% acetic acid for 15 min, washed three times in PBS for 5 min, dehydrated and rehydrated by sequential addition of 70%, 95%, 100%, 95%, and 70% EtOH, and washed three times in PBS. Cells were pre-hybridized for 4 h in HM buffer (50% Formamide, 5×SSC, 0.1% Tween 20, 9.2 mM citric acid, 50 μg/mL heparin, 500 μg/mL yeast RNA at pH 6.2) and hybridized overnight in 20 nM LNA probe in HM buffer in a humidified chamber at 55°C . Stringency washes were carried out in 100% HM, 50% HM/50% 2×SSC, twice in 2×SSC, and three times in 50% formamide/50% 2×SSC (each for 30 min) at hybridization temperature, followed by PBS, 0.1% Tween 20 at room temperature (five times for 5 min). Cells were blocked for 60 min in blocking buffer (2% sheep serum, 2 mg/mL BSA in PBS, 0.1% Tween 20) for 1 h at room temperature and stained with 1:500 FITC-conjugated streptavidin (Pierce) diluted in blocking buffer overnight at 4°C . Slides were washed three times in 2×SCC at 37°C , mounted, and visualized on a confocal microscope (Flowview 500; Olympus).

A Metamorph software package (Molecular Devices) was used to analyze the images. Concentric circles centered in the middle of the cell covering the cell body (diameter 70 μm, 140 μm, 210 μm, and 280 μm) were drawn and the intensity of the signals within each sector was measured. As expected, some low level signal was observed with the control and was subtracted. miR-26a extended well into tertiary and quaternary branches of the dendritic tree. All the dendrites contained signal with approximately equal intensity. In contrast, rno-miR-124a did not appreciably extend into the dendrites, especially relative to its signal in the soma. Consistent with their discrete localization in the RISC (RNA-induced silencing complex), dendritic miRNAs appeared punctate by in situ hybridization, a pattern typical of many dendritic mRNA populations. The visualization of miRNA puncta required capturing individual optical sections in the *Z*-axis.

We have also performed in situ hybridization on embryoid bodies from human embryonic stem cells grown on matrigel-coated 6-well plates or chamber slides (Lab-tech; Xu et al. 2009). The staining revealed the inverse expression patterns of embryonic stem cell markers and miR-145 expression. In this case, LNA probes were labeled with cy3-dUTP (Amersham) in 3′-End Labeling reaction (Roche). The slides were hybridized with LNA probes overnight and washed three times with 50% formamide/2× SSC and three times with 2× SSC at 45°C for 5 min each.

4.1 Profiling by deep sequencing

All profiling procedures so far described suffer from the problem that they can only detect known miRNAs. The growing availability of deep sequencing technologies will change this constraint dramatically. Depending on the preparatory procedure, in a single run, deep sequencing instruments will read all the small RNAs present in the sample, including as yet uncharacterized miRNAs and other categories of small RNAs. Quantification of the amount of each RNA present will remain relative, but the sensitivity and dynamic range are likely to rival or exceed quantitative PCR. When analyzed miRNAs have a characteristic appearance in which multiple reads sharply define the two ends of the mature miRNA, the absence of reads characterizes the loop region and a paucity of reads characterizes the opposite strand of the hairpin. In this rapidly moving field, the next step in the technology is the broad application of next-generation sequencing instrumentation to the many systems and questions now under study.

Acknowledgments We thank the W.M. Keck Foundation for supporting most of the work presented here.

References

Abu-Elneel K, Liu T, Gazzaniga FS, Nishimura Y, Wall DP, Geschwind DH, Lao K, Kosik KS (2008) Heterogeneous dysregulation of microRNAs across the autism spectrum. Neurogenetics 9:153–161

Bartel, DP (2004) MicroRNAs: genomics, biogenesis, mechanism and function. Cell 116:281–297

Chan JA, Krichevsky AM, Kosik KS (2005) MicroRNA-21 is an antiapoptotic factor in human glioblastoma cells. Cancer Res. 65:6029–6033

Krichevsky AM, King KS, Donahue CP, Khrapko K, Kosik KS (2003) A microRNA array reveals extensive regulation of microRNAs during brain development. RNA 9:1274–1281.

Krichevsky AM, Sonntag KC, Isacson O, Kosik KS (2006) Specific microRNAs modulate embryonic stem cell-derived neurogenesis. Stem Cells 24:857–864

Kye MJ, Liu T, Levy SF, Xu NL, Groves BB, Bonneau R, Lao K, Kosik KS (2007) Somatodendritic microRNAs identified by laser capture and multiplex RT-PCR. RNA 13:1224–1234

Lao K, Xu NL, Yeung V, Chen C, Livak KJ, Straus NA (2006) Multiplexing RT-PCR for the detection of multiple miRNA species in small samples. Biochem Biophys Res Commun. 343:85–89

Liu T, Papagiannakopoulos T, Puskar K, Qi S, Santiago F, Clay W, Lao K, Lee Y, Nelson SF, Kornblum HI, Doyle F, Petzold L, Shraiman B, Kosik KS (2007) Detection of a microRNA signal in an in vivo expression set of mRNA. PLoS ONE. 2:e804

Mercer TR, Dinger ME, Mariani J, Kosik KS, Mehler MF, Mattick JS (2008) Noncoding RNAs in long-term memory formation. Neuroscientist 14:434–445

Monticelli S, Ansel KM, Xiao C, Socci ND, Krichevsky AM, Thai TH, Rajewsky N, Marks DS, Sander C, Rajewsky K, Rao A, Kosik KS (2005) MicroRNA profiling of the murine hematopoietic system. Genome Biol. 6:R71

Papagiannakopoulos T, Shapiro A, Kosik KS (2008) MicroRNA-21 targets a network of key tumor-suppressive pathways in glioblastoma cells. Cancer Res 68:8164-8172

Stefani G, Slack FJ (2008) Small non-coding RNAs in animal development. Nature Rev Mol Cell Biol 9:219–230

Xu N, Papagiannakopoulos T, Pan G, Thomson JA, Kosik KS (2009) A repressive role for MicroRNA-145 on OCT4, SOX2 and KLF4 and human embryonic stem cell pluripotency. Cell 137:647–658

The Wide Variety of miRNA Expression Profiles in the Developing and Mature CNS

Marika Kapsimali

Abstract Although microRNA (miRNA) expression is prominent in the central nervous system (CNS), very few miRNAs have been studied in detail in this system. We performed a comprehensive analysis of the neuroanatomical expression profiles of 38 abundantly conserved miRNAs in developing and adult zebrafish brain. We observed a wide variety of different miRNA expression profiles in neural cells, ranging from single cell types to the majority of CNS cells and from transient to constitutive expression. The survey of miRNA expression patterns suggested several modes of action within neural cells, such as function in neural stem cells/progenitors, clearance of target mRNAs at spatial or temporal transitions, silencing of aberrant transcription and spatial and/or temporal regulation of target mRNA translation within mature neural cells.

1 Introduction

Among the hundreds of miRNAs found in vertebrate genomes (Bentwich et al. 2005; Berezikov et al. 2005; Cummins et al. 2006), a substantial number shows spatially and/or temporally restricted expression patterns in the CNS (Lagos-Quintana et al. 2002; Dostie et al. 2003; Sempere et al. 2004; Chen et al. 2005; Kloosterman et al. 2006b; Cheng et al. 2007; Hohjoh and Fukushima 2007). However, spatial and temporal expression profiles and functions of very few vertebrate miRNAs have been examined in detail. Within the vertebrate CNS, proposed roles for miRNAs include neurogenesis (Visvanathan et al. 2007), regulation of morphogenesis (Giraldez et al. 2005), dendrite formation (Schratt et al. 2006), and silencing of non-neural mRNAs (Smirnova et al. 2005; Vo et al. 2005; Conaco et al. 2006).

M. Kapsimali
INSERM U784, Génétique Moléculaire du Développement, Ecole Normale Supérieure, 46, rue d´ Ulm, 7530, Paris Cedex 05

B. de Strooper and Y. Christen (eds.), *Macro Roles for MicroRNAs in the Life and Death of Neurons*, Research and Perspectives in Neurosciences,
DOI 10.1007/978-3-642-04298-0_2,

miRNAs are also implicated in neurological diseases (Abelson et al. 2005; Chan et al. 2005; Ciafre et al. 2005; see also other chapters of this book). Although these studies point to the importance of miRNAs in brain development, function and disease, we still have little idea of the range of miRNA activities in neural cells. The detailed analysis of miRNA expression is a first step towards the discovery of their mode of action in the CNS.

To provide an overview of the breadth of miRNA expression in the CNS, we selected 38 vertebrate conserved miRNAs that had a particular and not ubiquitous expression profile in the CNS. For those 38 miRNAs, we performed in situ hybridizations with locked nucleic acid (LNA) probes and provided a comprehensive analysis of their neuroanatomical expression in developing and adult zebrafish brain (Kapsimali et al. 2007). Some of the miRNAs we analyzed belong to the same family or cluster and can differ in only one nucleotide located in, or outside, the "seed" sequence. To examine if the LNA probes can discriminate between miRNAs having only one or more different nucleotides, we performed in situ hybridization for some miRNAs using one or two internal mismatches. These experiments supported the conclusion that probes with two or more internal nucleotide differences detect signal from a single miRNA and not others with similar sequence (Kloosterman et al. 2006a; Kapsimali et al. 2007).

Here I will discuss some examples of the miRNA brain expression profiles that we found through systematic neuroanatomical analysis, in context with recent data about miRNA functions. To assist the reader, I will avoid the use of detailed neuroanatomical terms; a detailed description of miRNA expression in the developing and mature zebrafish nervous system can be found in Kapsimali et al. (2007).

2 miRNAs can be widely expressed in proliferating and/or differentiated cells throughout the brain

At larval (three and five days post-fertilization) stages, proliferative cells are present throughout the zebrafish brain, including the periventricular telencephalic, thalamic and hypothalamic zones, tectal proliferative zone, cerebellar valvula and rhombic lip (Wullimann and Knipp 2000). *miR-92b* is expressed in proliferative zones throughout the 5dpf larval zebrafish brain. This pattern indicates that *miR-92b* expression is present in most proliferative neural cells, irrespective of the fates of the progeny of these cells. In contrast, *miR-124*, *miR-138*, and other miRNAs are expressed in differentiating cells of the larval brain. Such patterns indicate that expression is associated with differentiation with little specificity regarding the identity of the differentiating neural cells.

In addition to miRNAs whose expression is restricted to either proliferating or differentiating cells, *miR-9* (Fig. 1A), *miR-135c*, *miR-153a*, *miR-219* and members of the *let-7* family show expression in both proliferating and differentiating cells of the larval brain. These patterns indicate that the expression of some miRNAs is not

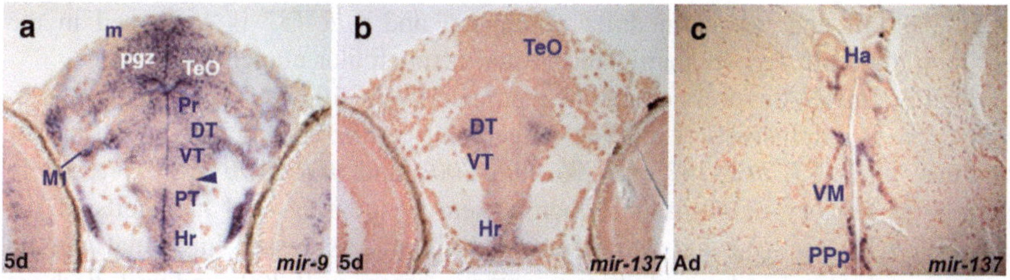

Fig. 1 Examples of brain miRNA expression. **A**. *miR-9* is broadly expressed in both proliferating (mainly periventricular) and differentiated CNS cells. Note the absence of expression in a "stripe"-like area between the posterior tuberculum (PT) and ventral thalamus (VT, arrowhead). **B, C**. *miR-137* expression is restricted and mainly observed in the nuclei of 5-day larval (**B**) and adult (Ad; **C**) zebrafish brain. Note the asymmetric expression in the habenular nuclei (Ha).
Abbreviations: DT: dorsal thalamus, Ha: habenular nuclei, Hr: rostral hypothalamus, m: tectal proliferative zone, M1: migrated pretectal area, pgz: periventricular gray zone of optic tectum, PPp: parvocellular preoptic nucleus-posterior part, Pr: periventricular pretectum, PT: posterior tuberculum, TeO: optic tectum, VM: ventromedial thalamic nucleus, VT: ventral thalamus

associated with a transition in the maturation state of the expressing cells. Among these miRNAs, *miR-9* has a striking expression profile. *miR-9* is expressed in telencephalic, diencephalic and tectal periventricular proliferative zones as well as in the mature neurons that arise from these domains, but it is downregulated or absent in "stripe"-forming cells (Fig. 1A).

3 miRNAs can show spatially restricted expression in the larval brain

In contrast to the miRNAs that are broadly expressed in proliferative and/or differentiated CNS cells, many others have larval expression that is restricted to specific brain areas, nuclei or cell types.

miR-222 and *miR-34* are expressed in neural cells in restricted subdivisions along the rostro-caudal axis of the larval brain. *miR-222* expression is restricted to specific groups of differentiating cells of the forebrain and midbrain, including telencephalon, eminentia thalami and hypothalamic nuclei. In contrast, *miR-34* expression is absent from forebrain and midbrain and present only in the caudal ventral and lateral isthmus and hindbrain nuclei, including the presumptive octaval area and the Mauthner cell.

miR-181a and *miR-181b* are strongly expressed in cells associated with the visual system, including retinal amacrine cells and ganglion cells, migrated pre-tectal and tectal cells. In addition, *miR-181a* and *miR-181b* are also expressed in the central pallium and medulla oblongata.

miR-128 and *miR-137* also have larval expression that is restricted to specific brain areas/nuclei. For example, *miR-128* is strongly expressed in the olfactory bulb, pallium and lateral medulla oblongata, and *miR-137* is expressed in the pallium, thalamic, hypothalamic and posterior tubercular nuclei (Fig. 1B).

Finally, miRNA expression can be restricted to specific cell types. For instance, *miR-218a* is exclusively expressed in cranial motor nuclei and spinal motor neurons; *miR-183, miR-182, miR-96, miR-200a* and *miR-200b* are expressed in peripheral sensory neuromasts, olfactory sensory neurons and hair cells of the ear. *miR-183, miR-182* and *miR-96* are additionally expressed in cranial ganglia neurons and retinal and pineal photoreceptors whereas *miR-200a* and *miR-200b* are expressed in taste buds. Finally, *miR-375* is expressed in the pituitary and a few scattered hypothalamic cells.

In conclusion, in the larval brain, miRNAs show widely divergent profiles of expression, varying from wide to very restricted expression either in particular brain subdivisions/areas or cell types.

4 Is miRNA expression conserved between larval and adult brain?

miRNAs preferentially target mRNAs with spatially or temporally complementary expression (Stark et al. 2005), raising the possibility that the miRNA requirement (and therefore expression) may be limited and coincide with spatial or temporal transitions of the target gene expression. Alternatively, miRNAs could maintain expression in the same cells or cell types throughout life and perhaps fulfil additional roles. We therefore examined the temporal regulation of miRNA expression between larval and adult stages to determine whether their expression is conserved throughout life.

In general, *miRNAs* with widespread expression in proliferating CNS cells and/or differentiated neurons maintain their expression from larval stages to the adult brain. For instance, *miR-92b* is expressed in periventricular cells and proliferative zones in the adult brain as it is in the larval brain. *miR-124* expression is excluded from periventricular cells and detected in most differentiated cells throughout the adult brain as in the case of the larval brain. Likewise, *miR-9* is expressed widely in periventricular zones and in many differentiated cells throughout the adult brain as in the larval brain.

miRNAs with spatially localized expression can also maintain their expression profile into adulthood. For instance, *miR-137* shows conserved expression in larval and adult brain in cells of the pallium, thalamus, ventral posterior tubercular area and other areas (Fig. 1B, C). *miR-222* shows conserved restricted expression in the rostral brain throughout life, with domains in the telencephalon, hypothalamus and posterior tubercular area.

These results show that, subsequent to their initial induction, some miRNAs conserve their expression in similar proliferating, differentiated or both cell groups throughout life. Although we cannot formally prove that expression is in same cells over time, our results strongly suggest that many miRNAs that are induced upon neuronal differentiation maintain constitutive expression throughout the life of the neurons.

In striking contrast to these brain miRNA profiles that are conserved throughout life, the expression of some miRNAs changes dramatically between larval and adult brain. For instance, in contrast to the strong and widespread expression of *miR-219* in some areas of the larval brain, including the dorsal thalamus, adult expression of *miR-219* is restricted to relatively few cells. Strikingly, in adults, *miR-219* is expressed in cells, possibly glia, associated with major tracts such as the lateral olfactory tract, and post-optic commissure/optic chiasm and tract, whereas the equivalent pathways in larvae are devoid of staining. These differences between larva and adult brain suggest downregulation of expression in most cells in the adult brain and either conserved or de novo expression in a few discrete cell populations. Such a loss of expression sites is consistent with functions of miRNAs in the regulation of genes that are only transcribed during restricted developmental phases.

Several miRNAs show de novo expression in adults that may reflect expression in late differentiating cell types not present or not fully differentiated in larval stages. For instance, *miR-34* shows conserved expression in the larval and adult ventral and lateral hindbrain, including the octavolateral nuclei, but also expands to include forebrain and midbrain cells, including cells of the habenulae, posterior tuberculum, pretectum, optic tectum as well as areas of the dorsal hindbrain such as the vagal lobe (the taste primary brain center of teleost fish). In a similar manner, *miR-218a* expression expands rostrally in the adult brain and it is observed not only in adult motor nuclei, as was the case in the larva, but also in telencephalic, hypothalamic and tectal areas.

Therefore, miRNA expression profiles can be conserved or divergent between larval and adult zebafish brain.

5 miRNAs of the same family or cluster can show subtle differences in brain expression

We compared the adult brain expression of miRNAs belonging to the same family, such as *miR-181a* and *miR-181b*, or cluster, such as *miR-221* and *miR-222*, that differ in three and four nucleotides, respectively; therefore, LNA probes should discriminate each of them. *miR-181a* and *miR-181b* show similar regional expression in the larval and adult brain, but in the adult brain we noticed differences in expression of *miR-181a* and *miR-181b* that were not obvious at larval stages. For instance, although both are expressed in the caudal hypothalamus, expression appears to be in different cells. One likely explanation for differences in expression is the genomic duplication of *miR-181a* and *miR-181b* and the different genomic localization of the additional copies (Hubbard et al. 2009).

miRNAs belonging to a particular cluster seem to largely share expression patterns but also have subtle differences in transcript localization. For instance, *miR-222* and *miR-221* share largely similar expression in the adult hypothalamus, but only *miR-222* is expressed in the ventral intermediate hypothalamus at the larval stage. It is not obvious why there should be differences in *miR-222* and *miR-221* expression, as they are present in the same cluster and one would predict that they are co-transcribed. This difference might be due to differential post-transcriptional regulation of miRNA expression (Obernosterer et al. 2006; Thomson et al. 2006).

6 What do the miRNA expression profiles predict about the role of miRNAs in the brain?

The aim of the neuroanatomical analysis of miRNA expression was to provide in a comprehensive way as much detail as possible about miRNAs expression in the CNS. Altogether this work provides a broad overview of miRNA expression in the brain and a foundation for future functional analyses. Although we cannot make specific conclusions regarding function based upon expression pattern alone, expression profiles do allow us to make generalized predictions regarding miRNA roles.

First, many miRNAs show expression associated with a transition in the differentiation state of the expressing cells. For instance, *miR-92b* is downregulated in most mature neurons. Conversely, *miR-124* is absent from proliferative cells and widely expressed in differentiated neurons. These patterns are consistent with miRNAs targeting genes expressed at different phases of differentiation. Several pieces of evidence supported the requirement of *miR-124* in neuronal differentiation, and more recently it was shown that *miR-124* controls neurogenesis in the adult subventricular zone by targeting repressors of neuronal differentiation such as sox9 (e.g., Conaco et al. 2006; Cheng et al. 2009).

Second, many of the expression profiles we describe are consistent with the miRNAs being expressed at high levels at times or places complementary to their targets. For instance, *miR-9* is broadly expressed in both proliferative and differentiated cells in many of its expression domains but avoids selected "stripes" of cells. These cells are located in areas such as the pallial/subpallial boundary, subcommissural organ, and zona limitans interthalamica, suggesting that *miR-9* is absent from areas with "organizing" activity in the brain. In this respect, the function of *miR-9* in the establishment of the mid-hindbrain boundary organizer was recently shown (Leucht et al. 2008).

Third, the constitutive expression of miRNAs such as *miR-124*, *miR-181*, *miR-222* and others in mature neurons is consistent with an initial role in the clearance of mRNAs from the neuronal precursor stage, but later they may fulfill a different role in the surveillance of fluctuations in aberrant transcription from notionally "silenced" loci. This finding is consistent with the idea that miRNAs confer robustness to programs of gene expression (Farh et al. 2005; Stark et al. 2005).

Fourth, one miRNA class was asymmetrically expressed in the zebrafish habenular nuclei (Fig. 1C), which is reminiscent of the well-studied role of miRNAs in the determination of left/right asymmetric fates for neurons in *C. elegans* (Chang et al. 2004), and it will be of interest to determine if vertebrate miRNAs are involved in the establishment of the robust asymmetries in neuronal organization of the habenulae and other brain areas.

Finally, miRNAs such as *miR-218a* are likely to be expressed in the same cells at the same time as their target genes. In such a situation, where miRNA and their target genes are co-expressed, miRNAs may spatially or temporally modulate protein levels within neural cells. At the moment, very little is known about miRNAs and protein level modulation in a given cell. A good example is *miR-134*, which is localized to dendrites and appears to modulate the levels of activity of a kinase that influences dendrite morphogenesis (Schratt et al. 2006).

7 Conclusion

In conclusion, this analysis reveals a wide diversity in miRNA expression, ranging from single cell types to the majority of CNS cells and from transient to constitutive expression. Altogether these data facilitate the prediction of roles that miRNAs may fulfill in the brain and provide a fundamental tool for miRNA functional analysis in the CNS.

Acknowledgments I am indebted to Steve Wilson, Ronald Plasterk, Wigard Kloosterman and Ewart de Bruijn for establishing this collaborative project. I am grateful to Frederic Rosa for all his support during this work and to the IPSEN foundation for giving me the opportunity to present our data in their conference. Finally, I would like to thank the funding organisms of this project (ZF-Models integrated project, CNRS, INSERM, Wellcome Trust, BBSRC, Council for Earth and Life Sciences from the Netherlands Organization for Scientific Research and ZonMw).

References

Abelson JF, Kwan KY, O'Roak BJ, Baek DY, Stillman AA, Morgan TM, Mathews CA, Pauls DL, Rasin MR, Gunel M, Davis NR, Ercan-Sencicek AG, Guez DH, Spertus JA, Leckman JF, Dure LSt, Kurlan R, Singer HS, Gilbert DL, Farhi A, Louvi A, Lifton RP, Sestan N, State MW (2005) Sequence variants in SLITRK1 are associated with Tourette's syndrome. Science 310:317–320

Bentwich I, Avniel A, Karov Y, Aharonov R, Gilad S, Barad O, Barzilai A, Einat P, Einav U, Meiri E, Sharon E, Spector Y, Bentwich Z (2005) Identification of hundreds of conserved and nonconserved human microRNAs. Nature Genet 37:766–770

Berezikov E, Guryev V, van de Belt J, Wienholds E, Plasterk RH, Cuppen E (2005) Phylogenetic shadowing and computational identification of human microRNA genes. Cell 120:21–24

Chan JA, Krichevsky AM, Kosik KS (2005) MicroRNA-21 is an antiapoptotic factor in human glioblastoma cells. Cancer Res 65:6029–6033

Chang S, Johnston RJ, Jr., Frokjaer-Jensen C, Lockery S, Hobert O (2004) MicroRNAs act sequentially and asymmetrically to control chemosensory laterality in the nematode. Nature 430:785–789

Chen PY, Manninga H, Slanchev K, Chien M, Russo JJ, Ju J, Sheridan R, John B, Marks DS, Gaidatzis D, Sander C, Zavolan M, Tuschl T (2005) The developmental miRNA profiles of zebrafish as determined by small RNA cloning. Genes Dev 19:1288–1293

Cheng HY, Papp JW, Varlamova O, Dziema H, Russell B, Curfman JP, Nakazawa T, Shimizu K, Okamura H, Impey S, Obrietan K (2007) microRNA modulation of circadian-clock period and entrainment. Neuron 54:813–829

Cheng LC, Pastrana E, Tavazoie M, Doetsch F (2009) miR-124 regulates adult neurogenesis in the subventricular zone stem cell niche. Nature Neurosci 12:399–408

Ciafre SA, Galardi S, Mangiola A, Ferracin M, Liu CG, Sabatino G, Negrini M, Maira G, Croce CM, Farace MG (2005) Extensive modulation of a set of microRNAs in primary glioblastoma. Biochem Biophys Res Commun 334:1351–1358

Conaco C, Otto S, Han JJ, Mandel G (2006) Reciprocal actions of REST and a microRNA promote neuronal identity. Proc Natl Acad Sci USA 103:2422–2427

Cummins JM, He Y, Leary RJ, Pagliarini R, Diaz LA, Jr., Sjoblom T, Barad O, Bentwich Z, Szafranska AE, Labourier E, Raymond CK, Roberts BS, Juhl H, Kinzler KW, Vogelstein B, Velculescu VE (2006) The colorectal microRNAome. Proc Natl Acad Sci USA 103:3687–3692

Dostie J, Mourelatos Z, Yang M, Sharma A, Dreyfuss G (2003) Numerous microRNPs in neuronal cells containing novel microRNAs. RNA 9:180–186

Farh KK, Grimson A, Jan C, Lewis BP, Johnston WK, Lim LP, Burge CB, Bartel DP (2005) The widespread impact of mammalian MicroRNAs on mRNA repression and evolution. Science 310:1817–1821

Giraldez AJ, Cinalli RM, Glasner ME, Enright AJ, Thomson JM, Baskerville S, Hammond SM, Bartel DP, Schier AF (2005) MicroRNAs regulate brain morphogenesis in zebrafish. Science 308:833–838

Hohjoh H, Fukushima T (2007) Expression profile analysis of microRNA (miRNA) in mouse central nervous system using a new miRNA detection system that examines hybridization signals at every step of washing. Gene 391:39–44

Hubbard TJ, Aken BL, Ayling S, Ballester B, Beal K, Bragin E, Brent S, Chen Y, Clapham P, Clarke L, Coates G, Fairley S, Fitzgerald S, Fernandez-Banet J, Gordon L, Graf S, Haider S, Hammond M, Holland R, Howe K, Jenkinson A, Johnson N, Kahari A, Keefe D, Keenan S, Kinsella R, Kokocinski F, Kulesha E, Lawson D, Longden I, Megy K, Meidl P, Overduin B, Parker A, Pritchard B, Rios D, Schuster M, Slater G, Smedley D, Spooner W, Spudich G, Trevanion S, Vilella A, Vogel J, White S, Wilder S, Zadissa A, Birney E, Cunningham F, Curwen V, Durbin R, Fernandez-Suarez XM, Herrero J, Kasprzyk A, Proctor G, Smith J, Searle S, Flicek P (2009) Ensembl 2009. Nucl Acids Res 37:D690–697

Kapsimali M, Kloosterman WP, de Bruijn E, Rosa F, Plasterk RH, Wilson SW (2007) MicroRNAs show a wide diversity of expression profiles in the developing and mature central nervous system. Genome Biol 8:R173

Kloosterman WP, Wienholds E, de Bruijn E, Kauppinen S, Plasterk RH (2006a) In situ detection of miRNAs in animal embryos using LNA-modified oligonucleotide probes. Nature Methods 3:27–29

Kloosterman WP, Steiner FA, Berezikov E, de Bruijn E, van de Belt J, Verheul M, Cuppen E, Plasterk RH (2006b) Cloning and expression of new microRNAs from zebrafish. Nucl Acids Res 34:2558–2569

Lagos-Quintana M, Rauhut R, Yalcin A, Meyer J, Lendeckel W, Tuschl T (2002) Identification of tissue-specific microRNAs from mouse. Curr Biol 12:735–739

Leucht C, Stigloher C, Wizenmann A, Klafke R, Folchert A, Bally-Cuif L (2008) MicroRNA-9 directs late organizer activity of the midbrain-hindbrain boundary. Nature Neurosci 11:641–648

Obernosterer G, Leuschner PJ, Alenius M, Martinez J (2006) Post-transcriptional regulation of microRNA expression. RNA 12:1161–1167

Schratt GM, Tuebing F, Nigh EA, Kane CG, Sabatini ME, Kiebler M, Greenberg ME (2006) A brain-specific microRNA regulates dendritic spine development. Nature 439:283–289

Sempere LF, Freemantle S, Pitha-Rowe I, Moss E, Dmitrovsky E, Ambros V (2004) Expression profiling of mammalian microRNAs uncovers a subset of brain-expressed microRNAs with possible roles in murine and human neuronal differentiation. Genome Biol 5:R13

Smirnova L, Grafe A, Seiler A, Schumacher S, Nitsch R, Wulczyn FG (2005) Regulation of miRNA expression during neural cell specification. Eur J Neurosci 21:1469–1477

Stark A, Brennecke J, Bushati N, Russell RB, Cohen SM (2005) Animal MicroRNAs confer robustness to gene expression and have a significant impact on 3'UTR evolution. Cell 123:1133–1146

Thomson JM, Newman M, Parker JS, Morin-Kensicki EM, Wright T, Hammond SM (2006) Extensive post-transcriptional regulation of microRNAs and its implications for cancer. Genes Dev 20:2202–2207

Visvanathan J, Lee S, Lee B, Lee JW, Lee SK (2007) The microRNA miR-124 antagonizes the anti-neural REST/SCP1 pathway during embryonic CNS development. Genes Dev 21:744–749

Vo N, Klein ME, Varlamova O, Keller DM, Yamamoto T, Goodman RH, Impey S (2005) A cAMP-response element binding protein-induced microRNA regulates neuronal morphogenesis. Proc Natl Acad Sci USA 102:16426–16431

Wullimann MF, Knipp S (2000) Proliferation pattern changes in the zebrafish brain from embryonic through early postembryonic stages. Anat Embryol 202:385–400

Viswanathan SR, Daley GQ, Gregory RI (2008) Selective blockade of microRNA processing by Lin28. Science 320:97–100
Wu L, Belasco JG (2008) Let me count the ways: mechanisms of gene regulation by miRNAs and siRNAs. Mol Cell 29:1–7

Interactions between microRNAs and Transcription Factors in the Development and Function of the Nervous System

David J. Simon

Abstract The formation and maturation of the nervous system requires precise control over the magnitude and timing of gene expression. MicroRNAs have emerged as potent regulators of translation, with roles ranging from the initial establishment of connectivity to activity-dependent refinement of the synapse. Many of the microRNA targets identified in the nervous system are themselves transcription factors, adding an additional layer of complexity in gene expression. Moreover, microRNAs and transcription factors are often embedded in feedback loops that reinforce changes in gene expression. These interactions lie at the heart of many cell fate decisions during the development of the nervous system. In addition, translational control of transcription factor abundance plays a prominent role in the regulation of retrograde signaling at neuromuscular synapses of both *C. elegans* and *Drosophila*.

1 Introduction

Precise translational control is critically important to the development and function of all tissues. Of particular interest is the nervous system whose unique morphology and need for plasticity places distinct demands on gene expression. In their role as potent regulators of translation, microRNAs are ideally suited to face these challenges. Though originally appreciated as developmental regulators (Reinhart et al. 2000; Wightman et al. 1993), microRNAs are clearly important in a number of biological processes, both in the nervous system and more broadly (Ambros 2004; Bartel 2004).

D.J. Simon
Department of Molecular Biology, Massachusetts General Hospital Boston, 185 Cambridge Street, Simches 7, Boston, MA02114, USA

B. de Strooper and Y. Christen (eds.), *Macro Roles for MicroRNAs in the Life and Death of Neurons*, Research and Perspectives in Neurosciences,
DOI 10.1007/978-3-642-04298-0_3,

The study of microRNAs in the nervous system is in its infancy; however, a number of prominent roles for microRNAs have already been established, from the earliest specification of the neuronal cell fate (Conaco et al. 2006) to activity-dependent refinement of synaptic structure (Schratt et al. 2006). Two general themes have emerged for the function of microRNAs in the nervous system: rheostats regulating the expression of a single gene, and switches regulating entire developmental programs. In their roles as rheostats, microRNAs control the protein abundance of one or more genes whose dosage is a critical determinant of a particular catalytic function. For example, translational control of Lim Kinase 1 by miR-134, a microRNA whose abundance is regulated by BDNF exposure, acutely modulates the morphology of mature dendritic spines (Schratt et al. 2006). In addition, miR-132, whose activity-dependent expression is governed by CREB, modulates neurite outgrowth through regulation of p250GAP expression (Vo et al. 2005). These and other well-established roles for microRNAs as regulators of gene expression have been extensively reviewed in the recent literature (Bicker and Schratt 2008).

In their roles as switches, microRNAs regulate a target whose dosage must pass a critical threshold in order to function in a specific process. Inherent in this idea is the notion of bistability, wherein subtle translational control of a target can have a disproportionate effect on its function within a cell. This control is often accomplished through the regulation of transcription factors, helping to explain the switch-like behavior of some microRNAs (Hobert 2006). These microRNA/transcription factor interactions can be broadly grouped into feedforward regulation, wherein a microRNA controls the translation of a transcription factor that goes on to activate or repress a set of genes, and more complicated feedback loops, wherein a transcription factor transcribes and is targeted by the same microRNA. The strongest evidence for both of these paradigms comes from the developing nervous system, where microRNA/transcription factor interactions are critical to the establishment of neuronal identity in several populations of cells.

Precise specification of sensory neurons in *Drosophila* requires the activation of pro-neural gene expression in sensory organ precursors and the lateral inhibition of these genes in neighboring cells through activation of the Notch pathway. This delicate balance is achieved by the dosage of the transcription factor *senseless.* Low expression of *senseless* represses transcription of pro-neural genes via Notch, whereas high levels directly activate pro-neural gene expression in sensory organ precursors. Li et al. (2006) have found that miR-9a functions in this pathway by directly inhibiting translation of *senseless.* There are more sensory organ precursor cells in both miR-9a mutants and gain-of-function mutations in *senseless*, whereas over-expression of miR-9a dramatically reduces the number of these cells. Importantly lowering *senseless* gene dosage in a miR-9a mutant background restores the number of sensory neurons to wild-type, providing strong evidence that miR-9a alters the function of the *senseless* protein to control the specification of sensory neuron number (Li et al. 2006). In the *Drosophila* maxillary palp, cell fate decisions regarding the number of sensory neurons that express receptors for carbon dioxide are made in part by the transcription factor Nerfin-1. Cayirlioglu and colleagues (2008) found that loss of miR-279 results in an increased number of these sensory neurons, a phenotype that is restored by lowering the expression of Nerfin-1, validating the regulation of Nerfin-1 levels by miR-279 as critical to the execution of a cell fate decision.

While direct translational control over transcription factor abundance can have enormous consequences for neuronal identity, microRNAs and transcription factors are often embedded in complicated feedback loops, allowing mutual regulation between these factors. The transcriptional repressor Yan is expressed in *Drosophila* eye progenitor cells and degraded following EGFR (epidermal growth factor receptor) signaling to enable maturation of these cells into photoreceptors. Yan functions in a negative feedback loop wherein it represses the transcription of miR-7, a microRNA that directly inhibits the translation of Yan. Over-expression of miR-7 promotes photoreceptor differentiation by lowering expression of Yan; however, loss of miR-7 does not have an obvious phenotype, suggesting that the microRNA works in concert with the normal EGFR-mediated degradation of Yan to reinforce rather than govern photoreceptor differentiation (Li and Carthew 2005).

Further evidence for a role of microRNA-transcription factor interactions in establishing cell fate decisions comes from the study of left/right asymmetric cell fate specification of developing *C.elegans* ASE chemosensory neurons. In these cells, the establishment of the left fate requires promotion of that fate by the DIE-1 transcription factor and inhibition of the right cell fate, which is achieved through the action of the COG-1 transcription factor. This asymmetry is achieved through a bistable feedback loop operating in both ASEL and ASER (ASE left/right). In ASEL, DIE-1 promotes the expression of the microRNA lsy-6, which inhibits expression of COG-1 and promotes ASEL fate. Conversely, in ASER, COG-1 activates expression of several members of the miR-273 family, which coordinately repress DIE-1 levels to prevent ASEL fate and promote ASER-specific gene expression (Johnston and Hobert 2003; Chang et al. 2004; Hobert 2006).

In vertebrates, the transcriptional repressor REST (repressor-element-1-silencing transcription factor) helps shape neuronal identity by preventing neuronal gene expression in non-neuronal cells. One of the genes inhibited by REST is the micro-RNA miR-124, a negative regulator of REST expression. This feedback loop reinforces neuronal gene expression by maintaining high levels of miR-124 in neurons and high levels of REST in non-neuronal cells (Conaco et al. 2006; Visvanathan et al. 2007). As the vertebrate nervous system develops further, the transcription factor Pitx3 promotes the dopaminergic cell fate among a population of midbrain cells. During this process, Pitx3 both activates expression of, and is inhibited by, the microRNA miR-133b, providing another example of a microRNA/transcription factor feedback loop at the heart of neuronal differentiation (Kim et al. 2007).

2 MicroRNA-Transcription Factor Interactions at the Synapse

While translational control by microRNAs can have a transformative effect on cell fate development in the nervous system, microRNA/transcription factor interactions also have a potent regulatory effect on the development and function of mature structures in the nervous system, including the synapse. Several recent examples at the neuromuscular junction (NMJ) highlight this emerging role in shaping neuronal connectivity and synaptic function.

In *Drosophila,* as with many other organisms, there is a selective pruning of synaptic connectivity during development. Mutants lacking the let-7 microRNA display fewer and smaller synaptic contacts during pupal development of abdominal musculature, a phenotype that is suppressed by lowering the dosage of the let-7 target *abrupt*, a transcription factor whose dosage has been established as a critical regulator of neuromuscular connectivity (Caygill and Johnston 2008; Hu et al. 1995). As development proceeds, *let-7* mutants fail to prune synaptic contacts with abdominal muscles so juvenile connectivity patterns persist through adulthood (Sokol et al. 2008). This phenotype, which is conceptually similar to the heterochronic phenotype of *C.elegans let-7* mutants (Reinhart et al. 2000), is predicted to arise through regulation of *abrupt*; however, experiments aimed at suppressing the *let-7* phenotype by lowering abrupt expression have not yet been performed. Interestingly, while let-7 is expressed in both muscle and neurons, *abrupt* expression is strongest in muscle, suggesting that the synaptic phenotype of *let-7* mutants might arise through a retrograde signal from muscle to neurons (Caygill and Johnston 2008; Sokol et al. 2008).

The magnitude of neurotransmitter release is often matched to the excitability of the post-synaptic membrane through the action of a retrograde signal (Davis and Bezprozvanny 2001). In *C.elegans*, this retrograde signal occurs when synaptic activity through the levamisole-sensitive nicotinic acetylcholine receptor (levR), one of two classes of nicotinic receptors in body-wall muscle, activates the activity-dependent transcription factor MEF-2, culminating in dampened pre-synaptic neurotransmitter release (Simon et al. 2008). Both of the critical components of this retrograde signaling pathway, the activity-sensing levR and the effector MEF-2, are targeted by the microRNA miR-1, such that modulation of miR-1 levels would change threshold for induction of retrograde signaling. Accordingly, retrograde signaling is constitutively active and neurotransmitter release is dampened in *mir-1* mutants. This phenotype can be suppressed by lowering the expression of MEF-2 in muscle, suggesting that a microRNA/transcription factor interaction lies at the heart of the coordination of pre- and post-synaptic function at the NMJ (Simon et al. 2008).

2.1 The mechanisms of microRNA/transcription factor interaction

While the consequences of microRNA-mediated repression of transcription factors can be dramatic, the mechanism underlying these phenotypes remains understudied.

It is temping to speculate that regulation of transcription factor abundance can affect transcriptional choice in a manner analogous to transcriptional co-factors and chromatin-modifying enzymes. This could be accomplished by directly changing the affinity for a particular consensus site or by enabling the transcription factor to become a *de novo* substrate for a signaling pathway. Amplified over the entire program of genes bound by a single transcription factor, regulation by a microRNA has the capacity to alter the transcriptional landscape of a given cell type.

To more fully understand the transcriptional consequences of microRNA/ transcription factor interactions, gene expression profiles in microRNA mutants will need to be systematically analyzed. Of interest is whether the genes differentially regulated in microRNA mutants are normally activated or repressed by the transcription factor in question. Consistent with this idea, a recent study in *Drosophila* used an allelic series of progressively severe mutations in D-MEF2 to define the temporal order of D-MEF2-dependent transcription. Interestingly, it found that genes that require higher levels of D-MEF2 are transcribed later during development, in agreement with known developmental increases in D-MEF2 levels (Elgar et al. 2008). When viewed through the lens of miR-1-mediated repression of MEF-2 (Simon et al. 2008), it is feasible that this temporal gradient of gene expression can be ascribed to the function of a single microRNA.

2.2 Further identification of microRNA-transcription factor interactions

Our understanding of the function of microRNAs in the nervous system is limited by the small number of individual microRNA knockouts and by our ability to define the full complement of microRNAs and their targets. In looking forward to exploring the full spectrum of microRNA/transcription factor interactions in the nervous system, it becomes important to consider how microRNAs regulate their targets.

Many microRNAs occur in large families grouped together by a similar seed sequence (Ambros et al. 2003; Grad et al. 2003; Lim et al. 2003). For example, the *let-7* family in *C.elegans* consists of four microRNAs with identical seed sequences and sparse complementarity at the 3' end. Successive deletion of these family members increases the severity of the heterochronic phenotypes of *let-7* single mutants, pointing to a synergistic regulation of target genes (Abbott et al. 2005). Consistent with this idea, a recent study generated knockouts of 83% of the known microRNAs in *C.elegans* and found that no single knockout generated a clear phenotype (Miska et al. 2007). Thus the potential exists for redundancy in the targeting of mRNAs by microRNA family members. Moreover, complex and cell type-specific regulation of gene structure must be taken into account. Splicing events often shape the size of the 3' untranslated region (UTR) of an mRNA, which would affect the number of microRNA binding sites. This is exemplified by the finding that, in cases where mRNAs are alternatively polyadenylated, the size of the 3'UTR can have a large impact on the number and diversity of microRNA binding sites (Legendre et al. 2006).

The discussion above focused on single targets of individual microRNAs. Yet there is emerging evidence to suggest that multiple microRNAs can regulate common targets and that any given microRNA may have tens if not hundreds of targets. Over-expression of individual microRNAs in HeLa cells led to down-regulation of hundreds of mRNAs, the majority of which contained seed region matches to the microRNA (Lim et al. 2005). This finding led to the notion that seed matches are the major determinant of microRNA binding. A subsequent search for evolutionarily conserved human 3'UTRs with perfect complementarity to the seed sequence of known microRNAs led to the assertion that upwards of one third of human genes are probable microRNA targets (Lewis et al. 2005). Further, proteomic analysis in over-expression and knock-down experiments points to a large number of mRNA targets for a given microRNA, ranging from subtle modulation to strong translational inhibition (Baek et al. 2008; Selbach et al. 2008).

A concern in interpreting over-expression studies is that the microRNA, whose base-pairing with mRNA targets is determined by the free energy of binding, might promiscuously regulate a target due to mass action. In fact, there are several examples of microRNAs with the bulk of their complementarity in the 3' region that efficiently down-regulate their targets (Brennecke et al. 2005). Central to this debate is the extent of microRNA:mRNA complementarity required and how distributed throughout the ~21 nucleotides the complementarity must be. At the extremes, it has been suggested that the only functional unit of the microRNA is the "seed" region, composed of nucleotides 2-7 (Lewis et al. 2005), whereas other groups have suggested that imperfect base pairing in the seed region can occur and complementarity with other regions of the microRNA is important (Didiano and Hobert 2006). This problem is only likely to be resolved as more loss of function mutations in microRNA genes become available to establish a set of universal rules for target prediction.

3 Concluding Remarks

By regulating other factors involved in gene expression, microRNAs can have a transformative effect on the identity and function not only in cells where they are expressed but also in neighboring cells through the regulation of cell-cell signaling. While this discussion has focused largely on the regulation of transcription factor abundance by microRNAs in the nervous system, many important transcription factors in a variety of cell types are also targeted by microRNAs, including GATA-1 (erythropoiesis), SOX4 and c-myc (cell fate and differentiation), c-Myb (lymphocyte development) and E2F1 (cell cycle progression) (Chang et al. 2008; Dore et al. 2008; O'Donnell et al. 2005; Tavazoie et al. 2008; Xiao et al. 2007). As our understanding of the full repertoire of microRNAs/transcription factor interactions grows, so too will our appreciation of their diverse regulatory functions.

Acknowledgments I thank Professor Joshua Kaplan for critical reading of this manuscript and for his support during my time in his laboratory.

References

Abbott AL, Alvarez-Saavedra E, Miska EA, Lau NC, Bartel DP, Horvitz HR, Ambros V (2005) The let-7 MicroRNA family members mir-48, mir-84, and mir-241 function together to regulate developmental timing in Caenorhabditis elegans. Dev Cell 9:403–414

Ambros V, Lee RC, Lavanway A, Williams PT, Jewell D (2003) MicroRNAs and other tiny endogenous RNAs in C. elegans. Curr Biol 13:807–818

Ambros V (2004) The functions of animal microRNAs. Nature 431:350–355

Baek D, Villén J, Shin C, Camargo FD, Gygi SP, Bartel DP (2008) The impact of microRNAs on protein output. Nature 455:64–71

Bartel DP (2004) MicroRNAs: Genomics biogenesis mechanism and function. Cell 116:281–297

Bicker S, Schratt G (2008) microRNAs: tiny regulators of synapse function in development and disease. J Cell Mol Med 12:1466–1476

Brennecke J, Stark A, Russell RB, Cohen S M (2005) Principles of microRNA-target recognition. PLoS Biol 3: e85

Caygill EE, Johnston LA (2008) Temporal regulation of metamorphic processes in Drosophila by the let-7 and miR-125 heterochronic microRNAs. Curr Biol 18:943–950

Cayirlioglu P, Kadow IG Zhan X, Okamura K, Suh GS, Gunning D, Lai EC, Zipursky SL (2004) Hybrid neurons in a microRNA mutant are putative evolutionary intermediates in insect CO2 sensory systems. Science 319:1256–60

Chang S, Johnston RJ, Frokjaer-Jensen C, Lockery S,Hobert O (2004) MicroRNAs act sequentially and asymmetrically to control chemosensory laterality in the nematode. Nature 430:785–789

Chang T-C, Yu D, Lee Y-S, Wentzel EA, Arking DE, West KM, Dang CV, Thomas-Tikhonenko A, Mendell J T (2008) Widespread microRNA repression by Myc contributes to tumorigenesis. Nature Genet 40:43–50

Conaco C, Otto S, Han J-J, Mandel G (2006) Reciprocal actions of REST and a microRNA promote neuronal identity. Proc Natl Acad Sci USA 103:2422–2427

Davis GW, Bezprozvanny I (2001) Maintaining the stability of neural function: a homeostatic hypothesis. Ann Rev Physiol 63:847–869

Didiano D, Hobert O (2006) Perfect seed pairing is not a generally reliable predictor for miRNA-target interactions. Nature Struct Mol Biol 13:849–851

Dore LC, Amigo JD, dos Santos CO, Zhang Z, Gai X, Tobias JW, Yu D, Klein AM, Dorman C, Wu W Hardison RC, Paw BH, Weiss MJ (2008) A GATA-1-regulated microRNA locus essential for erythropoiesis. Proc Natl Acad Sci USA 105:3333–3338

Elgar SJ, Han J, Taylor MV (2008) mef2 activity levels differentially affect gene expression during Drosophila muscle development. Proc Natl Acad Sci USA 105:918–923

Grad Y, Aach J, Hayes GD, Reinhart BJ, Church GM, Ruvkun G, Kim J (2003) Computational and experimental identification of C elegans microRNAs. Mol Cell 11:1253–1263

Hobert O (2006) Architecture of a MicroRNA-controlled gene regulatory network that diversifies neuronal cell fates Cold Spring Harbor Symp Quant Biol 71:181–188

Hu S, Fambrough D, Atashi JR, Goodman CS, Crews ST (1995) The Drosophila abrupt gene encodes a BTB-zinc finger regulatory protein that controls the specificity of neuromuscular connections, Genes Dev 9:2936–48

Johnston RJ, Hobert O (2003) A microRNA controlling left/right neuronal asymmetry in Caenorhabditis elegans. Nature 426:845–849

Kim J, Inoue K, Ishii J, Vanti WB, Voronov SV, Murchison E, Hannon G, Abeliovich A (2007) A microRNA feedback circuit in midbrain dopamine neurons. Science 317:1220–1224

Legendre M, Ritchie W, Lopez F, Gautheret D (2006) Differential repression of alternative transcripts: a screen for miRNA targets. PLoS Comput Biol 2:e43

Lewis BP, Burge CB, Bartel DP (2005) Conserved seed Pairing, often flanked by adenosines, indicates that thousands of human genes are microRNA targets Cell 120:15–20

Li X, Carthew RW (2005) A microRNA mediates EGF receptor signaling and promotes photoreceptor differentiation in the Drosophila eye. Cell 123:1267–1277

Li Y, Wang F, Lee JA, Gao FB (2006) MicroRNA-9a ensures the precise specification of sensory organ precursors in Drosophila. Genes Dev 20:2793–2805

Lim LP, Lau NC, Garrett-Engele P, Grimson A, Schelter JM, Castle J, Bartel DP, Linsley PS, Johnson JM (2005) Microarray analysis shows that some microRNAs downregulate large numbers of target mRNAs. Nature 433:769–773

Lim LP, Lau NC, Weinstein EG, Abdelhakim A, Yekta S Rhoades MW, Burge CB, Bartel DP (2003) The microRNAs of Caenorhabditis elegans. Genes Dev 17:991–1008

Miska EA, Alvarez-Saavedra E, Abbott AL, Lau NC, Hellman AB, McGonagle SM, Bartel DP, Ambros VR, Horvitz HR (2007) Most Caenorhabditis elegans microRNAs are individually not essential for development or viability PLoS Genetics 3:e215

O'Donnell KA, Wentzel EA, Zeller KI, Dang CV, Mendell JT (2005) c-Myc-regulated microRNAs modulate E2F1 expression. Nature 435:839–843

Reinhart BJ, Slack FJ, Basson M, Pasquinelli AE, Bettinger JC, Rougvie AE, Horvitz HR, Ruvkun G (2000) The 21-nucleotide let-7 RNA regulates developmental timing in Caenorhabditis elegans. Nature 403:901–906

Schratt GM, Tuebing F, Nigh EA, Kane CG, Sabatini ME, Kiebler M, Greenberg ME (2006) A brain-specific microRNA regulates dendritic spine development Nature 439:283–289

Selbach M, Schwanhäusser B, Thierfelder N, Fang Z, Khanin R, Rajewsky N (2008) Widespread changes in protein synthesis induced by microRNAs. Nature 455:58–63

Simon DJ, Madison JM, Conery AL, Thompson-Peer KL, Soskis M, Ruvkun GB, Kaplan JM, Kim JK (2008) The microRNA miR-1 regulates a MEF-2-dependent retrograde signal at neuromuscular junctions. Cell 133:903–915

Sokol NS, Xu P, Jan YN, Ambros V (2008) Drosophila let-7 microRNA is required for remodeling of the neuromusculature during metamorphosis, Genes Dev 22:1591–1596

Tavazoie SF, Alarcon C, Oskarsson T, Padua D, Wang Q, Bos PD, Gerald WL, Massague J (2008) Endogenous human microRNAs that suppress breast cancer metastasis. Nature 451:147–152

Visvanathan J, Lee S, Lee B, Lee JW, Lee S-K (2007) The microRNA miR-124 antagonizes the anti-neural REST/SCP1 pathway during embryonic CNS development. Genes Dev 21:744–749

Vo N, Klein ME Varlamova O, Keller DM, Yamamoto T, Goodman RH, Impey S (2005) A cAMP-response element binding protein-induced microRNA regulates neuronal morphogenesis. Proc Natl Acad Sci USA 102:16426–16431

Wightman B, Ha I, Ruvkun G (1993) Posttranscriptional regulation of the heterochronic gene lin-14 by lin-4 mediates temporal pattern formation in C elegans. Cell 75:855–862

Xiao C, Calado DP, Galler G, Thai T-H, Patterson HC, Wang J, Rajewsky N, Bender TP, Rajewsky K (2007) MiR-150 controls B cell differentiation by targeting the transcription factor c-Myb. Cell 131:146–159

A microRNA Feedback Circuit in Midbrain Dopamine Neurons

Asa Abeliovich

Abstract Midbrain dopamine neurons (mDNs) play a central role in complex behaviors such as reward and addiction, and these cells are lost in Parkinson's disease (PD). A number of transcription factors have been implicated in the regulation of mDNs. However, the role of post-transcriptional mechanisms in mDNs, or in other post-mitotic neuron types, is relatively uncharacterized. Here we investigate the role of microRNAs (miRNAs) in mDN regulation in relation to previously described transcriptional control mechanisms. miRNAs are evolutionarily conserved, 18-25 nucleotide, non-protein coding transcripts that play an important function in post-transcriptional regulation of gene expression during development. In preliminary studies and a recent manuscript, we identified eight miRNA enriched in the midbrain, and one – miR-133b – as a mDN-enriched miRNA that appears to function within a feedback circuit with the homeodomain transcription factor Pitx3.

miRNAs are derived from long primary transcripts through sequential processing by the Drosha ribonuclease (Lee et al. 2003) and the Dicer enzyme (Grishok et al. 2001; He and Hannon 2004). In the context of an RNA-induced silencing complex (RISC), miRNAs guide the cleavage of target mRNAs and/or inhibit their translation (Meister and Tuschl 2004). miRNAs were first characterized in invertebrates, where they function to regulate developmental cell fate decisions in the nervous system (Chang et al. 2004; Johnston and Hobert 2003) and elsewhere (Ambros 2003). In vertebrates, several hundred miRNAs have been identified, but only a few of these have been associated with specific cellular functions. For instance, the mir-430 family of miRNAs appears to be required for normal brain morphogenesis in zebrafish development (Giraldez et al. 2005, 2006). Several miRNAs have been identified that are expressed during differentiation to a neuronal phenotype either

A. Abeliovich
Asa Abeliovich Columbia University Medical Center, 630 West 168th Street, Room 15-05, NY 10032, New York, USA
e-mail: aa900@columbia.edu

B. de Strooper and Y. Christen (eds.), *Macro Roles for MicroRNAs in the Life and Death of Neurons*, Research and Perspectives in Neurosciences,
DOI 10.1007/978-3-642-04298-0_4,

in vivo (Kosik and Krichevsky 2005; Krichevsky et al. 2003) or in the context of embryonic stem (ES) cell cultures (Krichevsky et al. 2006). More recently, a mammalian miRNA was identified that regulates neurite morphology in embryonic hippocampal neuron cultures (Schratt et al. 2006).

Embryonic and neuronal stem cells are capable of differentiating to the mDN phenotype in vitro and thus offer a potentially simplified model system for the analysis of mDN maturation relative to the intact mammalian CNS. Furthermore, stem-cell derived mDNs may be of therapeutic benefit in cell replacement strategies for neurodegenerative disorders because of their potentially limitless supply. A number of studies have shown that ES cultures can generate cells with a mDN phenotype and that the developmental program in vitro appears to recapitulate the temporal course of normal mDN maturation (Barberi et al. 2003; Kim et al. 2002; Martinat et al. 2006; Sonntag et al. 2004). Two predominant methods have been established for ES cell development: 1) an embryoid body- (EB)based protocol (Lee et al. 2000) that involves detachment of ES cells from an adherent surface, growth as cell aggregates in suspension, and subsequent reattachment; and 2) a co-culture-based protocol that uses bone marrow-derived stromal cells (Kawasaki et al. 2000). Both appear to generate cells that phenotypically mimic mDNs and express midbrain-specific phenotypic markers. Additional protocols that simplify the generation of mDNs have also been established (Andersson et al. 2006).

Soluble factors implicated in the specification and differentiation of mDNs, such as SHH, FGF8, BDNF, and GDNF (Barberi et al. 2003), further promote mDN development in ES cultures in vitro. Similarly, overexpression of cell-intrinsic transcription factors, such as Nurr1 (Kim et al. 2002; Sonntag et al. 2004), Pitx3 (Chung et al. 2005), and Lmx1a (Andersson et al. 2006), appear to induce the development of mDN phenotype in these cultures, consistent with their proposed in vivo functions. Thus, ES cell-based assays offer a potentially simplified model system in which to dissect the interactions and kinetics of cell intrinsic and extrinsic signals that govern mDN specification.

Importantly, ES cell-derived mDN appear to be relatively inefficient as replacement cells when tested in animal models of PD. For instance, rats that have been lesioned unilaterally within the substantia nigra with a mDN toxin, 6-hydroxydopamine, and are therefore deficient in striatal dopaminergic innervation on one side, display a stereotyped turning behavior in response to dopaminergic agonists such as apomorphine. Transplantation of ES-derived cells directly into the striatal target region can suppress this turning behavior (Kim et al. 2002), but only a small number of TH-positive cells are found within the transplants (Kim et al. 2002; Martinat et al. 2006). It is likely that the vast majority of transplanted cells do not survive, and those that do survive display reduced dopaminergic neuronal markers. These findings may reflect limited or ineffective development of the transplanted cells that therefore limits their survival, a lack of extrinsic support in the host adult striatum, or the presence of inhibitory factors. Most studies have used murine ES cells, but similar results have been obtained with human ES cultures (Ben-Hur et al. 2004; Martinat et al. 2006). Further understanding of endogenous mDN development and ES cell biology is necessary to address these limitations.

We sought to establish a role for miRNAs in mammalian dopamine neuron differentiation, function, and survival. To facilitate a kinetic analysis, we first used an in vitro model system: the differentiation of murine ES cells into DNs (Kim et al. 2002; Martinat et al. 2006). An ES cell line was obtained that conditionally expresses Dicer enzyme by harboring LoxP recombinase sites that flank both chromosomal copies of the Dicer gene (floxed Dicer alleles; Murchison et al. 2005). Introduction of Cre recombinase into these cells by lentiviral transduction leads to the deletion of Dicer in nearly 100% of cells (data not shown). Floxed Dicer ES cultures were differentiated to a mDN phenotype using the EB protocol (Martinat et al. 2006). Briefly, cells were initially grown in non-adherent conditions in the context of defined media, including growth factors, to generate neuronal precursors (stage 2); subsequently, neuronal precursors were expanded in the presence of basic fibroblast growth factor (bFGF; stages 3 and 4); and finally, the bFGF was withdrawn to obtain mature DNs (stage 5), which constitute 10–25% of the cells in these cultures as quantified by the expression of markers including tyrosine hydroxylase (TH).

Cre-mediated deletion of Dicer at stage 4, when postmitotic dopamine neurons first arise, led to a nearly complete loss of dopamine neuron accumulation at stage 5. Similarly, other neuronal classes, including GABAergic neurons, were reduced in these cultures, although to a lesser degree. Cre lentivirus had no effect on wild-type cells (data not shown). The number of cells expressing TujI, an early general neuronal marker that first appears at the neural precursor stage (stage 4) of EB differentiation, was reduced by approximately 50%. The reduction in dopamine neuron accumulation appeared to be the result of both increased neuronal apoptosis, as demonstrated by the appearance of apoptotic markers including activated caspase-3 and TUNEL staining, and reduced neurogenesis, as quantified by BrdU incorporation. The Dicer deletion phenotype is significantly rescued by transfection of low molecular weight RNA species derived from embryonic mouse midbrain, consistent with a model in which miRNA play a role in midbrain dopamine neuron terminal differentiation and survival.

In a second approach, mice were generated that are homozygous for the floxed Dicer allele and express Cre recombinase under the regulation of dopamine transporter regulatory sequences (DAT:CRE$^{+/-}$; Dicer$^{flox/flox}$; (Murchison et al. 2005)), thus leading to the specific deletion of Dicer in CNS dopamine neurons. These mice display a progressive loss of mDNs that is first apparent at two weeks of age and is nearly complete by six weeks of age. Taken together, these results indicate that Dicer is essential for the terminal differentiation and survival of multiple neuron types, including mDNs.

A limitation to standard approaches for the purification of miRNAs from mDNs is that these cells typically represent a minor population within tissue or primary culture preparations. To circumvent this problem, we took a subtractive approach and compared miRNA expression profiles of normal adult midbrain with the profiles of midbrain depleted of dopamine neurons. We used three independent models to this end: mice treated with the dopamine neuron-specific toxin 6-hydroxydopamine (6-OHDA; Martinat et al. 2006); adult *Aphakia* mice deficient in the transcription factor Pitx3 (Hwang et al. 2003; Nunes et al. 2003; van den Munckhof et al. 2003); and adult human Parkinson's disease (PD) patients. These models were validated by real-time quantitative PCR (qPCR) on cDNA preparations, demonstrating a dramatic reduction in mDN marker expression.

Expression analyses were performed by qPCR for a panel of approximately 224 miRNAs precursors in midbrain, cerebellum, and cerebral cortex samples from PD patients and normal controls. Within this miRNA panel, eight appeared to be specifically expressed in midbrain relative to cerebral cortex or cerebellum. Expression of one of these miRNA, mir-133b, was specifically deficient in the context of PD patient samples, as confirmed by Northern blotting for mir-133b. Expression of several other miRNAs within the panel was reduced in PD patients relative to normal controls, but expression of these miRNAs was not specific to the midbrain (data not shown). Additional miRNAs within the panel were found to be specifically enriched in the midbrain, cerebellum or cerebral cortex but were not altered in the PD patient samples.

The relatively complete deficiency of mir-133b expression in Aphakia midbrain was surprising, as adult *Aphakia* mice do maintain a population of midbrain dopamine neurons within the ventral tegmental area (VTA; data not shown; Nunes et al. 2003); this finding suggested the possibility that mir-133b is a direct target of Pitx3 transcription activation (as Pitx3 is mutated in Aphakia mice). Consistent with this model, overexpression of Pitx3 in stage 4 murine ES cultures leads to a dramatic up-regulation of mir-133b precursor expression. Alignment of the proximal promoter regions of Mir-133b and TH in search of conserved regulatory cis-acting elements identified a short sequence that corresponds precisely to a Pitx3 binding site. Expression of a luciferase reporter gene that harbors approximately 350 bp of proximal miR-133b promoter sequences is specifically induced by over-expression of Pitx3 in COS cells. Mir-133b expression within murine ES differentiation cultures appears bimodal, with an initial expression at stage 3 and subsequent induction at stage 5, closely mimicking the Pitx3 expression pattern. Taken together, these data suggest that Pitx3 regulates mir-133b expression in mDNs.

Next, we sought to investigate the function of mir-133b in mDN maturation, function, and survival. A lentiviral vector was generated that harbors miR-133b precursor sequences. Transduction of COS cells with this vector, but not a control vector (miR-18 or GFP), inhibits expression of a luciferase reporter that harbors a precise target sequence for mir-133b within its 3′ untranslated sequences. We investigated the consequences of increased miR-133b expression in either ES-derived cultures or in primary embryonic midbrain cultures. miR-133b precursor overexpression induced a relative reduction in the expression of mDN maturation markers, including TH, and DAT, although transcription of early mDN markers, such as Pitx3 and Nurr1, appeared unaltered. Consistent with this finding, dopamine release in the context of potassium-induced depolarization was markedly reduced in the context of miR-133b overexpression in either ES cultures or in primary midbrain cultures. Overexpression of miR-133b at the neural precursor stage of EB-differentiated ES cultures led to a significant reduction in the number of

TH-positive cells and to a lesser reduction of TuJI-positive cells in the ES cultures. In the primary midbrain cultures, which harbor post-mitotic mDNs, overexpression of mir-133b did not significantly alter the number of TH-positive cells (data not shown); thus mir-133 does not appear to directly modify the survival of mDNs in these cultures.

Next, we investigated the effect of suppressing mir-133b in the context of mDNs. The activity of mir-133b could be inhibited using a 2′OMe-modified RNA oligonucleotide homologous to the miR-133b sequence and linked to a short peptide derived from the Drosophila Antennapedia protein that mediates cell transduction (Davidson et al. 2004; Schratt et al. 2006). qPCR indicated increased gene expression of TH and DAT in the context of reduced mir-133b activity in primary rat midbrain cultures (data not shown). Consistent with this finding, the mir-133b inhibitory oligonucleotide induced a significant increase in potassium-stimulated dopamine release in both the ES-derived and the primary midbrain cultures. Finally, inhibition of mir-133b at stage 4 of EB differentiation led to a small apparent increase in the number of TH+ cells, but this increase did not reach statistical significance. Taken together, these data implicate mir-133b in the regulation of dopamine neuron function and terminal differentiation.

Individual microRNAs appear to regulate numerous targets post-translationally. To identify potential physiological targets for miR-133b activity, we used several available miRNA target prediction programs based on 3′ untranslated sequence homology to mir133b (John et al. 2004; Lewis et al. 2003). Interestingly, the Pitx3 3'-untranslated region (UTR) was found to be a potential target of mir-133b activity. As Pitx3 is thought to play a critical role in the regulation mDN gene expression (Martinat et al. 2006; Smidt et al. 2004), a hypothetical model for the observed phenotypes associated with altered mir-133b expression is that miR-133b normally suppresses Pitx3 expression, leading secondarily to reduced mDN marker transcription. This model is consistent with the finding that targets of Pitx3 activity, including TH and DAT, are modified in transcription in the context of miR-133b mis-expression. To test whether Pitx3 3′ UTR sequences are subject to direct regulation by mir-133b activity in vivo, we placed these sequences downstream of a luciferase reporter gene. The putative mir-133b target site in the Pitx3 3′ UTR led to a reduction in luciferase accumulation in a mir-133b-dependent manner. Fluorescence activated cell sorter (FACS) analysis of primary rat midbrain cultures with an antibody for Pitx3 revealed that mir-133b overexpression induced a reduction in Pitx3 protein in both TH+ and TH− cell populations, and miR-133b knockdown led to an increase in Pitx3 protein. Of note, Pitx3 mRNA transcript levels did not appear to be altered by mir-133b manipulation, consistent with a post-transcriptional mechanism.

If Pitx3 is a significant direct target of mir-133b, one prediction is that overexpression of a Pitx3 transgene lacking 3′-UTR regulatory elements would "rescue" the reduced mDN marker expression phenotype associated with mir-133b overexpression. Consistent with this prediction, mir-133b-induced reduction in TH and DAT transcription can be suppressed by overexpression of Pitx3 lacking the 3′UTR miR-133b target sequences. Also consistent with this model, mir-133b inhibition fails to induce TH and DAT expression in the absence of Pitx3. Finally, we find that in miRNA-deficent, Dicer mutant TH^{+} mDNs derived from $DAT{:}CRE^{+/-}$; $Dicer^{flox/flox}$ mice, Pitx3 protein expression is upregulated relative to control (DAT: $CRE^{+/-}$; $Dicer^{flox/+}$) cells, consistent with a role for miRNA in Pitx3 regulation.

References

Ambros V (2003) MicroRNA pathways in flies and worms: growth, death, fat, stress, and timing. Cell 113:673–676

Andersson E, Tryggvason U, Deng Q, Friling S, Alekseenko Z, Robert B, Perlmann T, Ericson J (2006) Identification of intrinsic determinants of midbrain dopamine neurons. Cell 124:393–405

Barberi T, Klivenyi P, Calingasan NY, Lee H, Kawamata H, Loonam K, Perrier AL, Bruses J, Rubio ME, Topf N, Tabar V, Harrison NL, Beal MF, Moore MA, Studer L (2003) Neural subtype specification of fertilization and nuclear transfer embryonic stem cells and application in parkinsonian mice. Nature Biotechnol 21:1200–1207

Ben-Hur T, Idelson M, Khaner H, Pera M, Reinhartz E, Itzik A, Reubinoff BE (2004) Transplantation of human embryonic stem cell-derived neural progenitors improves behavioral deficit in Parkinsonian rats. Stem Cells 22:1246–1255

Chang S, Johnston RJ Jr., Frokjaer-Jensen C, Lockery S, Hobert O (2004) MicroRNAs act sequentially and asymmetrically to control chemosensory laterality in the nematode. Nature 430:785–789

Chung S, Hedlund E, Hwang M, Kim DW, Shin BS, Hwang DY, Jung Kang U, Isacson O, Kim KS (2005) The homeodomain transcription factor Pitx3 facilitates differentiation of mouse embryonic stem cells into AHD2-expressing dopaminergic neurons. Mol Cell Neurosci 28:241–252

Davidson TJ, Harel S, Arboleda VA, Prunell GF, Shelanski ML, Greene LA, Troy CM (2004) Highly efficient small interfering RNA delivery to primary mammalian neurons induces MicroRNA-like effects before mRNA degradation. J Neurosci 24:10040–10046

Giraldez AJ, Cinalli RM, Glasner ME, Enright AJ, Thomson JM, Baskerville S, Hammond SM, Bartel DP, Schier AF (2005) MicroRNAs regulate brain morphogenesis in zebrafish. Science 308:833–838

Giraldez AJ, Mishima Y, Rihel J, Grocock RJ, Van Dongen S, Inoue K, Enright AJ, Schier AF (2006) Zebrafish MiR-430 promotes deadenylation and clearance of maternal mRNAs. Science 312:75–79

Grishok A, Pasquinelli AE, Conte D, Li N, Parrish S, Ha I, Baillie DL, Fire A, Ruvkun G, Mello CC (2001) Genes and mechanisms related to RNA interference regulate expression of the small temporal RNAs that control C. elegans developmental timing. Cell 106:23–34

He L, Hannon GJ (2004) MicroRNAs: small RNAs with a big role in gene regulation. Nature Rev Genet 5:522–531

Hwang DY, Ardayfio P, Kang UJ, Semina EV, Kim KS (2003) Selective loss of dopaminergic neurons in the substantia nigra of Pitx3-deficient aphakia mice. Brain Res Mol Brain Res 114:123–131

John B, Enright AJ, Aravin A, Tuschl T, Sander C, Marks DS (2004) Human MicroRNA targets. PLoS Biol 2z; e363

Johnston RJ, Hobert O (2003) A microRNA controlling left/right neuronal asymmetry in Caenorhabditis elegans. Nature 426:845–849

Kawasaki H, Mizuseki K, Nishikawa S, Kaneko S, Kuwana Y, Nakanishi S, Nishikawa SI, Sasai Y (2000) Induction of midbrain dopaminergic neurons from ES cells by stromal cell-derived inducing activity. Neuron 28:31–40

Kim JH, Auerbach JM, Rodriguez-Gomez JA, Velasco I, Gavin D, Lumelsky N, Lee SH, Nguyen J, Sanchez-Pernaute R, Bankiewicz K, McKay R (2002) Dopamine neurons derived from embryonic stem cells function in an animal model of Parkinson's disease. Nature 418:50–56

Kosik KS, Krichevsky AM (2005) The elegance of the microRNAs: a neuronal perspective. Neuron 47:779–782

Krichevsky AM, King KS, Donahue CP, Khrapko K, Kosik KS (2003) A microRNA array reveals extensive regulation of microRNAs during brain development. RNA 9:1274–1281

Krichevsky AM, Sonntag KC, Isacson O, Kosik KS (2006) Specific microRNAs modulate embryonic stem cell-derived neurogenesis. Stem Cells 24:857–864

Lee SH, Lumelsky N, Studer L, Auerbach JM, McKay RD (2000) Efficient generation of midbrain and hindbrain neurons from mouse embryonic stem cells. Nature Biotechnol 18:675–679

Lee Y, Ahn C, Han J, Choi H, Kim J, Yim J, Lee J, Provost P, Rådmark O, Kim S, Kim VN (2003) The nuclear RNase III Drosha initiates microRNA processing. Nature 425:415–419

Lewis BP, Shih IH, Jones-Rhoades MW, Bartel DP, Burge CB (2003) Prediction of mammalian microRNA targets. Cell 115:787–798

Martinat C, Bacci JJ, Leete T, Kim J, Vanti WB, Newman AH, Cha JH, Gether U, Wang H, Abeliovich A (2006) Cooperative transcription activation by Nurr1 and Pitx3 induces embryonic stem cell maturation to the midbrain dopamine neuron phenotype. Proc Natl Acad Sci USA 103:2874–2879

Meister G, Tuschl T (2004) Mechanisms of gene silencing by double-stranded RNA. Nature 431:343–349

Murchison EP, Partridge JF, Tam, OH, Cheloufi S, Hannon GJ (2005) Characterization of Dicer-deficient murine embryonic stem cells. Proc Natl Acad Sci USA 102:12135–12140

Nunes I, Tovmasian LT, Silva RM, Burke RE, Goff SP (2003) Pitx3 is required for development of substantia nigra dopaminergic neurons. Proc Natl Acad Sci USA 100:4245–4250

Schratt GM, Tuebing F, Nigh EA, Kane CG, Sabatini ME, Kiebler M, Greenberg ME (2006) A brain-specific microRNA regulates dendritic spine development. Nature 439:283–289

Smidt MP, Smits SM, Burbach JP (2004) Homeobox gene Pitx3 and its role in the development of dopamine neurons of the substantia nigra. Cell Tissue Res 318:35–43

Sonntag KC, Simantov R, Kim KS, Isacson O (2004) Temporally induced Nurr1 can induce a non-neuronal dopaminergic cell type in embryonic stem cell differentiation. Eur J Neurosci 19:1141–1152

van den Munckhof P, Luk KC, Ste-Marie L, Montgomery J, Blanchet PJ, Sadikot AF, Drouin J (2003) Pitx3 is required for motor activity and for survival of a subset of midbrain dopaminergic neurons. Development 130:2535–2542

Fine-tuning mRNA Translation at Synapses with microRNAs

Gerhard M. Schratt

Abstract Multiple activity-dependent programs of gene expression orchestrate the development and function of neural circuitries. The local synthesis of proteins in neuronal dendrites is critical for enduring forms of synaptic plasticity that might be a cellular substrate for information storage. Here, I will summarize our recent findings that implicate microRNAs (miRNAs) in the control of dendritic protein synthesis and synaptic plasticity in mammalian neurons. Large-scale functional screening indicates that at least three different miRNAs are involved in the morphological plasticity of dendritic spines, the specialized dendritic structures of synaptic contact. Detailed mechanistic studies revealed that miRNA-mediated regulation of dendritic proteins synthesis impinges on critical modulators of the actin cytoskeleton within spines. Dendritic miRNAs are subject to activity-dependent regulation at both the transcriptional and post-transcriptional levels, implicating miRNAs in the experience-dependent remodeling of neural circuits. By coordinating the global and local fine-tuning of proteins synthesis in response to activity, miRNAs might play a pivotal role in the maintenance of cellular and network homeostasis. Aberrant miRNA expression and/or function could underlie neuropsychiatric disorders that are characterized by defective homeostasis.

1 Introduction

Within the mammalian brain, neurons form intricate networks that underlie the ability of the brain to perform higher cognitive functions, such as learning and memory. On the other hand, malfunctioning of neural circuits is a frequent cause of

G.M. Schratt
Interdisziplinäres Zentrum für Neurowissenschaften, SFB488 Junior Group, Universität Heidelberg, and Institut für Neuroanatomie, Universitätsklinikum Heidelberg, Im Neuenheimer Feld 345, 69120, Heidelberg, Germany
e-mail: schratt@ana.uni-heidelberg.de

B. de Strooper and Y. Christen (eds.), *Macro Roles for MicroRNAs in the Life and Death of Neurons*, Research and Perspectives in Neurosciences,
DOI 10.1007/978-3-642-04298-0_5, © Springer-Verlag Berlin Heidelberg 2010

neurological disorders, including mental retardation, mood disorders and addiction. Neurons within circuits communicate with each other via synapses, which are specialized contact sites where electric stimulation is converted into biochemical signals. Typical excitatory synapses form on dendritic spines, the specialized protrusions from dendrites that harbor the postsynaptic neurotransmitter receptors and signaling machinery (Hering and Sheng 2001; Bonhoeffer and Yuste 2002). During early stages of postnatal development, the initial formation of synaptic contacts is governed by intrinsic genetic programs that are largely not influenced by neuronal activity (Waites et al. 2005). However, at later stages, activity-dependent programs of gene expression orchestrate the remodeling of neural circuits in response to experience. The developmental plasticity of neural circuits occurs at multiple levels, including the regulation of dendritic arborization, the functional and morphological remodeling of dendritic spines and juxtaposed presynaptic terminals, and the adjustment of the overall excitability of individual neurons. Moreover, plasticity of neural circuits persists into adulthood, where it mainly serves the purpose of encoding memory traces. Seminal work during the last two decades has documented the pivotal role of new gene expression for most forms of plasticity in vertebrate neurons (Flavell and Greenberg 2008). In particular, de novo synthesis of proteins from preexisting mRNAs has been shown to be required for the long-lasting changes in synaptic transmission that are believed to underlie learning and memory (Sajikumar et al. 2005). The synthesis of a limited set of proteins occurs locally within dendrites, and this local form of mRNA translation enables neurons to tightly regulate the buildup of crucial regulatory and structural proteins at the resolution of individual dendrites and possibly spines (Steward and Schuman 2001). While the importance of local mRNA translation for the morphological plasticity of dendrites and spines is now well documented, the underlying molecular mechanisms are less understood. Analogous to other systems in which local translation plays a role, e.g., oocyte development and fibroblast migration, ribonucleoprotein particles (RNPs) inhibit premature translation of dendritic mRNAs during transport and storage at synaptic sites (Kiebler and DesGroseillers 2000). Major components of such RNPs are sequence-specific RNA-binding proteins and regulatory RNAs, which are mostly recruited to the 3′ untranslated region (UTR) of the mRNA to exert their function. However, the exact composition of these RNPs, as well as their dynamic regulation in response to extracellular cues (i.e., synaptic stimulation), remains to be determined. I will try to summarize our recent findings that implicate microRNAs (miRNAs) in the regulation of local dendritic mRNA translation and synaptic development. miRNAs are an extensive class of small non-coding RNAs that act as post-transcriptional regulators of gene expression in a variety of biological and pathological processes (Kloosterman and Plasterk 2006). Recently, mammalian neurons were shown to contain an extensive repertoire of miRNAs that function at all stages of development (Kosik 2006; Fiore et al. 2008). Ultimately, it is our hope that knowledge about the function and mechanism of action of miRNAs during dendritic and synaptic plasticity might provide us with additional tools for therapeutic intervention in cognitive disorders.

2 Results

2.1 The prototypic dendritic microRNA: miR-134

In a search for molecules that could participate in the translational control of dendritic mRNAs upon brain-derived neurotrophic factor (BDNF) stimulation, we identified miR-134, a brain-specific member of the miRNA family that was recently cloned by the Tuschl lab (Lagos-Quintana et al. 2002). Within neuronal dendrites,

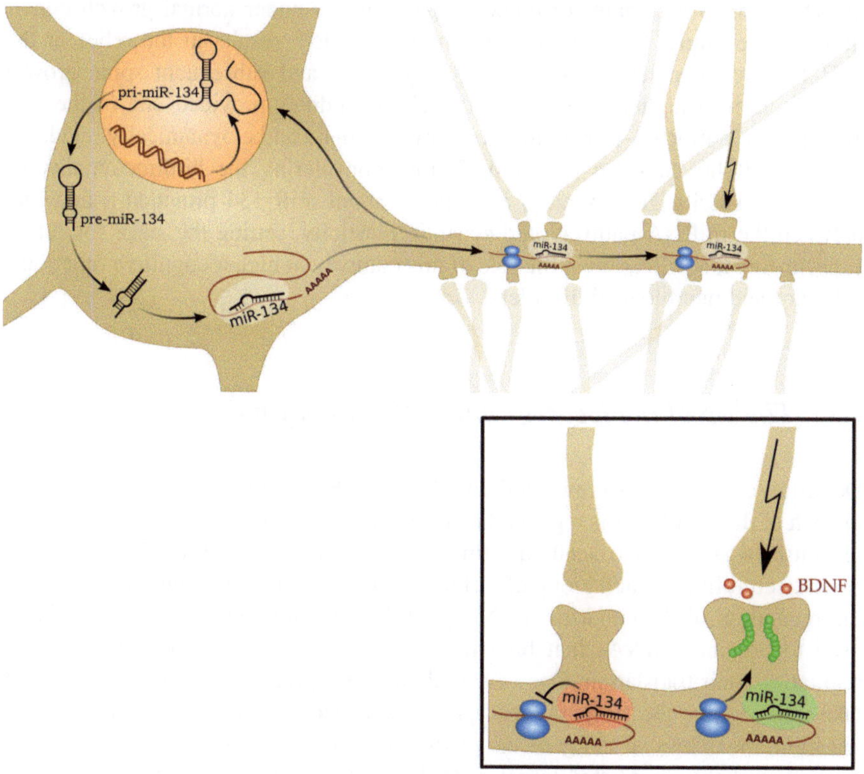

Fig. 1 miR-134-dependent control of dendritic spine size.
Transcription of the miR-134 gene yields a primary miR-134 transcript (pri-miR-134) that is processed in the nucleus to a precursor miR-134 hairpin RNA (pre-miR-134). Pre-miR-134 is exported in the neuronal soma, where it undergoes a second round of processing, leading to the mature miR-134. miR-134 is incorporated into a miRNA-associated silencing complex (miRISC), which is transported into the synaptodendritic compartment, presumably already in conjunction with dendritic mRNA targets (i.e., Limk1). During transport and storage at non-stimulated synapses, miR-134 blocks the translation of the associated mRNAs, thereby helping to restrict dendritic spine growth. Upon synaptic stimulation (flash), BDNF release triggers the inactivation of the miRISC, thereby inducing mRNA translation and spine growth (inset, below). Moreover, synaptic activity elicits new pri-miR-134 transcription, thereby globally reinforcing the silencing effect of miR-134. This feedback mechanism might have implications for the control of synapse homeostasis (see text for details)

miR-134, presumably in conjunction with the miRNA-associated, RNAi-induced silencing complex, binds to the 3'UTR of multiple mRNAs that encode for important synaptic proteins, including the actin regulator Lim domain-containing protein kinase 1 (Limk1; Fig.1; Schratt et al. 2006). Recruitment of the miR-134-associated translational silencing complex to the Limk1 mRNA results in the inhibition of Limk1 synthesis. Importantly, the miR-134-dependent regulation of Limk1 synthesis also occurs locally in dendrites, as evidenced by the use of a dendritically localized, diffusion-limited myrGFP-Limk1-3'UTR reporter construct. Ectopic expression of Limk1 can rescue the miR-134 spine phenotype, strongly suggesting that miR-134-dependent repression of Limk1 synthesis is required to restrict the growth of dendritic spines of hippocampal neurons under normal growth conditions. Interestingly, treatment of neurons with BDNF results in the relief of the miR-134-mediated repression of Limk1 synthesis and subsequent spine growth (Fig. 1, inset at higher magnification). These findings provided one of the first examples of a miRNA-target mRNA interaction that can be dynamically regulated by extracellular cues, an important feature considering the bi-directionality of synaptic plasticity. In summary, our studies about miR-134 provided a first link between the miRNA pathway and synaptic plasticity, setting the stage for future experiments that will address the role of this pathway in higher cognitive functions, behavior and neurological disorders.

2.2 *The tip of the iceberg: miR-138 and beyond*

Dendrites contain up to a few hundred different mRNA species, and the translation of each of these mRNAs is regulated in a very specific temporal and local manner in response to specific extracellular stimuli (Eberwine et al. 2001). Therefore, we considered it likely that a group of different dendritic miRNAs might participate in the regulation of the dendritic mRNA pool. In principle, bioinformatics could be used to predict miRNAs that bind to the 3′UTR of dendritic mRNAs, thereby regulating their translation. One caveat about these predictions is that they do not take into account the temporal and spatial expression domains of the miRNAs in question, meaning that many of the predictions could be false positives, since the miRNA and their predicted target mRNA are not co-expressed in dendrites. To circumvent this problem, we decided to define the complement of miRNAs expressed in the synaptodendritic compartment by using a comparative, genome-wide expression profiling approach. Thereby, we identified a set of 10 different miRNA species that are significantly and reproducibly enriched in rat brain synaptosomes, a biochemical preparation of highly purified pinched-off synaptic terminals (Siegel et al. 2009). This number is likely an underestimation, since it is derived by a very stringent selection procedure. Nevertheless, it provided us with a list of candidate synaptic miRNAs that could be tested in follow-up functional assays. For example, using dendritic spine size as our morphological readout, we found that loss-of-function for two of the candidate miRNAs, miR-132 and miR-138,

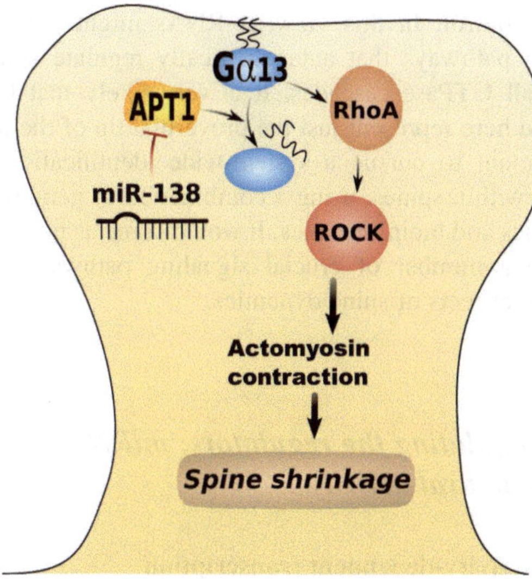

Fig. 2 miR-138-dependent control of dendritic spine size. Dendritically localized miR-138 inhibits translation of the mRNA encoding for acyl-protein thioesterase 1 (APT1), an enzyme catalyzing the removal of palmitate moieties from synaptic proteins, i.e., Gα13. The resulting increase in membrane-associated Gα13 results in hyperactivation of the RhoA signaling pathway, actomyosin contraction and spine shrinkage. How miR-138 function is modulated by synaptic activity is presently unclear

resulted in impaired dendritic spine morphogenesis. Notably, miR-138 was identified as a very potent inhibitor of dendritic spine size in cultured hippocampal neurons. miR-132, on the other hand, displayed a spine growth-promoting effect, in agreement with a previous study from the Goodman lab that investigated the role of miR-132 in dendritic outgrowth (Wayman et al. 2008). In contrast to miR-132, the downstream pathway of miR-138 in neurons was completely unknown. Therefore, we decided to investigate the underlying molecular mechanism of the miR-138 spine effect in more detail. Using a combination of bioinformatics target prediction and reporter gene assays, we identified the depalmitoylating enzyme acyl-protein thioesterase-1 (APT1) as a bona fide miR-138 target mRNA (Fig. 2). miR-138 inhibits production of the endogenous APT1 enzyme in hippocampal neurons. Importantly, miR-138-dependent downregulation of APT1 expression is required to restrict spine growth, implicating APT1, and depalmitoylation in general, in the control of dendritic spine structure. Spine shrinkage upon miR-138 overexpression correlates with a reduced amplitude of miniature excitatory postsynaptic currents (mEPSC), demonstrating for the first time that miRNAs regulate the synaptic transmission process. Work from the El-Husseini lab had shown that a huge number of important synaptic regulatory proteins are post-translationally modified by palmitoylation, which in turn controls membrane localization and activity of these proteins (Kang et al. 2008). One of the critical substrates of APT1 in the context of spine morphogenesis appears to be Gα12/13, an activator of the Rho signaling cascade (Kurose 2003). miR-138 increases Gα12/13 membrane localization, and membrane-attached, palmitoylated Gα12/13 is able to revert the spine-growth promoting effect of miR-138 inhibition.

From this and our previous studies, a common theme is emerging that the synthesis of critical regulators of the actin cytoskeleton within spines is under miRNA control. In this view, miRNAs might contribute to the fine-tuning of signaling pathways that antagonistically regulate spine growth and contraction, e.g., small GTPase cascades. It is very likely that the three dendritic miRNAs discussed here represent just the proverbial tip of the iceberg. In the future, it will be important to pursue a system-wide identification of the miRNA regulatory network within spines, using a combination of genome-wide expression profiling, proteomics and bioinformatics. It would come as no surprise if miRNAs turn out to fine-tune a number of crucial signaling pathways involved in functional and structural aspects of spine dynamics.

2.3 *Regulating the regulators: miRNAs under the control of neural activity*

2.3.1 Activity-dependent transcription

The remodeling of neural circuits during development and in the adult is under the control of activity patterns driven by diverse forms of experience (e.g., visual experience). Therefore, in addition to the knowledge of the miRNA-dependent pathways involved in plasticity, it is imperative to understand how miRNA expression itself is regulated by plasticity-inducing cues. A flurry of studies has highlighted the importance of activity-dependent transcription in neural plasticity (Hong et al. 2005). We therefore explored the effect of activity on the expression of dendritic miRNAs, initially focusing on miR-134. In developing hippocampal cultures, miR-134 expression gradually increases, with the highest expression observed between two and three weeks of culture (Schratt et al. 2006). Increased miR-134 expression coincides with elevated intrinsic activity within the cultures, prompting us to explore the effect of stimuli that mimic synaptic activation. Indeed, membrane-depolarizing concentrations of extracellular potassium (to trigger massive calcium influx) or BDNF treatment led to a robust and transient induction of miR-134 expression (Fiore et al. 2009). This activity-dependent miR-134 expression occurs primarily at the level of pri-miR-134 transcription. Strikingly, the miR-134 gene is embedded within a large cluster of >40 different miRNAs, and expression of the entire cluster is coordinately induced by stimuli that mimic neuronal activity, suggesting that these miRNAs could derive from a common, long precursor transcript. Using comparative genomics followed by chromatin immunoprecipitation, we found that the transcription factor MEF2 is bound to native chromatin upstream of the miRNA cluster. MEF2, which was previously shown to limit synapse formation in hippocampal cultures (Flavell et al. 2006; Shalizi et al. 2006), is necessary and sufficient for activity-induced expression of the miRNA cluster. Functionally, several members of the cluster, including miR-134, are necessary for an activity-dependent increase in dendritic complexity, suggesting that these miRNAs might regulate a set of target mRNAs that encode for regulators of dendrite growth. An important target in this respect appears to be Pumilio 2 (Pum2), since Pum2 is sufficient to rescue the defect in dendritogenesis observed upon miR-134 inhibition. Both increasing and decreasing Pum2 protein levels compromise dendritic growth, suggesting that fine-tuning Pum2 levels within a narrow window (e.g., by the control of miR-134) is important for this phenotype. In conclusion, miR-134, in addition to its inhibitory role in spine growth, fulfills a growth-promoting function during activity-dependent dendritogenesis, possibly in conjunction with other miRNAs from the cluster. We speculate that these apparently opposing functions of miR-134 could play a role in homeostatic adaptations within neural circuits.

In contrast to miR-134, expression of miR-138 precursors is not induced but rather repressed by increasing neuronal activity (Siegel et al. 2009). It is presently unclear whether this repression is due to reduced transcription of the pri-miR-138 or increased processing of pri-/pre-miR-138 in response to activity. The latter possibility is especially appealing, given that pre-miR-138 processing is blocked in most cell types by virtue of an unknown repressor molecule (Obernosterer et al. 2006). To further complicate matters, mature miR-138 is derived from two individual genomic loci, and thus the relative contribution of the different pri-miR-138 transcripts to the generation of mature miR-138 in neurons under basal and activity conditions will need to be determined. In conclusion, additional experiments are required to explore the relevance of activity-dependent miR-138 expression for the dynamic remodeling of dendritic spines.

2.3.2 Activity-dependent regulation of miRNA-associated complexes

Our findings that BDNF was able to release miR134 inhibition of Limk1 mRNA translation provided one of the first examples that the activity of mature miRNAs can be regulated by extrinsic stimuli (Fig. 1). Preliminary results indicate that miR-134 is still associated with the Limk1 mRNA upon BDNF stimulation. Therefore, we envision that the activation of intracellular kinase signaling cascades in response to BDNF (i.e., mTOR signaling) triggers post-translational modification(s) of key components of the miR-134-associated silencing complex (miRISC). This hypothesis is supported by the recent observations of Ago2 phosphorylation by MAPK, calpain-mediated Dicer cleavage and memory-induced degradation of the RISC helicase Armitage (Lugli et al. 2005; Ashraf et al. 2006; Zeng et al. 2008). In addition to these core components of RISC, sequence-specific RNA-binding proteins associated with the Limk1 3′UTR could modulate miR-134 activity. A detailed understanding of the dynamic nature of miRNA function will require a biochemical characterization of the miR-134-associated miRNP and the post-translational modifications of individual components.

Using a cleavage-based sensor assay, we found that miR-138 activity is also subject to activity-dependent regulation. However, in contrast to miR-134, increased neural activity (caused by membrane depolarization) reduces miR-138 cleavage activity towards the sensor, suggesting that activity triggers inactivation of the miR-138 inhibitory complex. The exact mechanism and localization of this regulatory event remain to be determined.

2.4 *Local vs. global: are miRNAs involved in synaptic homeostasis?*

While our work clearly demonstrates that the activity of synaptic miRNAs, such as miR-134 and miR-138, is regulated by multiple activity-dependent mechanisms, we do not yet understand how these pathways are coordinated in time and space within individual neurons. Concerning miR-134, it appears counterintuitive at first glance that one miRNA permits dendritic outgrowth in response to an increase in neural activity but restricts the growth of dendritic spines in the very same dendrites. However, recent evidence indicates that increased dendritic arborization and reduction in synaptic strength (downscaling) could be a coordinated response of neurons to prevent overexcitation during neural circuit development (Peng et al. 2009). In that sense, miR-134 could be an integral part of a feedback mechanism that keeps the global excitability of neurons within a narrow range (Turrigiano 2007). On top of this global function, miR-134 regulation possibly plays a role in the local control of protein synthesis, at the resolution of dendritic segments or even individual spines, in response to synaptic stimulation (Fig. 1). Such a mechanism could endow individual synapses with the capability to match their response to the strength of synaptic inputs they receive, despite a global up- or downscaling of synaptic weights due to changes in overall excitability. To address this homeostasis hypothesis experimentally, we plan to pursue two lines of investigation. First, we will explore the role of miR-134 in global synaptic scaling in response to chronic activation/activity blockade by measuring morphological and electrophysiological parameters. Second, we will monitor local changes in miR-134 activity in response to the stimulation of individual spines (BDNF puffing, glutamate uncaging) by monitoring the translation of dendritically localized, miR-134-dependent reporter genes. Overall, we hope to gain some fundamental insight into the mechanisms by which individual neurons coordinate local, synapse-specific Hebbian plasticity with global homeostatic plasticity. This knowledge will have important implications for the understanding of activity-dependent neural circuit development, memory storage and neuropsychiatric disorders characterized by impaired neuronal homeostasis (Ramocki and Zoghbi 2008).

3 Concluding remarks

A number of studies, including from our lab, have established miRNA-dependent control of gene expression as an important additional layer of regulation during dendrite and synapse development in primary neuron cultures. Preliminary experiments further indicate that miRNA expression and function are under the control of neuronal activity, suggesting that miRNAs could play a role in the experience-dependent plasticity of neural circuits. A major future challenge will be to test whether these in vitro findings translate into critical functions of miRNAs in vertebrate cognition. Finally, expression studies indicate a frequent deregulation of miRNA expression in a number of neurological disorders. The identification of individual miRNAs that are causally involved in the etiology of such diseases will be required to devise novel miRNA-based strategies for therapeutic intervention.

Acknowledgments I thank G. Siegel for help with the illustrations, R. Fiore for critically reading the manuscript and all members of my laboratory for stimulating discussions. Work in my laboratory is funded by the Deutsche Forschungsgemeinschaft (DFG, SFB488), the Human Frontier Science Program (HFSP, CDA) and the National Institute on Drug Abuse (NIDA, 1R21DA025102-01).

References

Ashraf SI, McLoon AL, Sclarsic SM, Kunes S (2006) Synaptic protein synthesis associated with memory is regulated by the RISC pathway in Drosophila. Cell 124:191–205

Bonhoeffer T, Yuste R (2002) Spine motility. Phenomenology, mechanisms, and function. Neuron 35:1019–1027

Eberwine J, Miyashiro K, Kacharmina JE, Job C (2001) Local translation of classes of mRNAs that are targeted to neuronal dendrites. Proc Natl Acad Sci USA 98:7080–7085

Fiore R, Siegel G, Schratt G (2008) MicroRNA function in neuronal development, plasticity and disease. Biochim Biophys Acta 1779:471–478

Fiore R, Khudayberdiev S, Christensen M, Siegel G, Flavell SW, Kim TK, Greenberg ME, Schratt G (2009) Mef2-mediated transcription of the miR379-410 cluster regulates activity-dependent dendritogenesis by fine-tuning Pumilio2 protein levels. Embo J 28:697–710

Flavell SW, Greenberg ME (2008) Signaling mechanisms linking neuronal activity to gene expression and plasticity of the nervous system. Annu Rev Neurosci 31:563–590

Flavell SW, Cowan CW, Kim TK, Greer PL, Lin Y, Paradis S, Griffith EC, Hu LS, Chen C, Greenberg ME (2006) Activity-dependent regulation of MEF2 transcription factors suppresses excitatory synapse number. Science 311:1008–1012

Hering H, Sheng M (2001) Dendritic spines: structure, dynamics and regulation. Nature Rev Neurosci 2:880–888

Hong EJ, West AE, Greenberg ME (2005) Transcriptional control of cognitive development. Curr Opin Neurobiol 15:21–28

Kang R, Wan J, Arstikaitis P, Takahashi H, Huang K, Bailey AO, Thompson JX, Roth AF, Drisdel RC, Mastro R, Green WN, Yates JR, 3rd, Davis NG, El-Husseini A (2008) Neural palmitoyl-proteomics reveals dynamic synaptic palmitoylation. Nature 456:904–909

Kiebler MA, DesGroseillers L (2000) Molecular insights into mRNA transport and local translation in the mammalian nervous system. Neuron 25:19–28

Kloosterman WP, Plasterk RH (2006) The diverse functions of microRNAs in animal development and disease. Dev Cell 11:441–450

Kosik KS (2006) The neuronal microRNA system. Nature Rev Neurosci 7:911–920

Kurose H (2003) Galpha12 and Galpha13 as key regulatory mediator in signal transduction. Life Sci 74:155–161

Lagos-Quintana M, Rauhut R, Yalcin A, Meyer J, Lendeckel W, Tuschl T (2002) Identification of tissue-specific microRNAs from mouse. Curr Biol 12:735–739

Lugli G, Larson J, Martone ME, Jones Y, Smalheiser NR (2005) Dicer and eIF2c are enriched at postsynaptic densities in adult mouse brain and are modified by neuronal activity in a calpain-dependent manner. J Neurochem 94:896–905

Obernosterer G, Leuschner PJ, Alenius M, Martinez J (2006) Post-transcriptional regulation of microRNA expression. RNA 12:1161–1167

Peng YR, He S, Marie H, Zeng SY, Ma J, Tan ZJ, Lee SY, Malenka RC, Yu X (2009) Coordinated changes in dendritic arborization and synaptic strength during neural circuit development. Neuron 61:71–84

Ramocki MB, Zoghbi HY (2008) Failure of neuronal homeostasis results in common neuropsychiatric phenotypes. Nature 455:912–918

Sajikumar S, Navakkode S, Frey JU (2005) Protein synthesis-dependent long-term functional plasticity: methods and techniques. Curr Opin Neurobiol 15:607–613

Schratt GM, Tuebing F, Nigh EA, Kane CG, Sabatini ME, Kiebler M, Greenberg ME (2006) A brain-specific microRNA regulates dendritic spine development. Nature 439:283–289

Shalizi A, Gaudilliere B, Yuan Z, Stegmuller J, Shirogane T, Ge Q, Tan Y, Schulman B, Harper JW, Bonni A (2006) A calcium-regulated MEF2 sumoylation switch controls postsynaptic differentiation. Science 311:1012–1017

Siegel G, Obernosterer G, Fiore R, Oehmen M, Bicker S, Christensen M, Khudayberdiev S, Leuschner PJ, Busch CL, Kane CG, Hübel K, Dekker F, Rengarajan B, Drepper C, Waldmann H, Kauppinen S, Greenberg ME, Draguhn A, Rehmsmeier M, Martinez J, Schratt G (2009) A functional screen implicates microRNA-138-dependent regulation of the depalmitoylation enzyme APT1 in dendritic spine morphogenesis. Nature Cell Biol 11:705–716

Steward O, Schuman EM (2001) Protein synthesis at synaptic sites on dendrites. Annu Rev Neurosci 24:299–325

Turrigiano G (2007) Homeostatic signaling: the positive side of negative feedback. Curr Opin Neurobiol 17:318–324

Waites CL, Craig AM, Garner CC (2005) Mechanisms of vertebrate synaptogenesis. Annu Rev Neurosci 28:251–274

Wayman GA, Davare M, Ando H, Fortin D, Varlamova O, Cheng HY, Marks D, Obrietan K, Soderling TR, Goodman RH, Impey S (2008) An activity-regulated microRNA controls dendritic plasticity by down-regulating p250GAP. Proc Natl Acad Sci USA 105:9093–9098

Zeng Y, Sankala H, Zhang X, Graves PR (2008) Phosphorylation of Argonaute 2 at serine-387 facilitates its localization to processing bodies. Biochem J 413:429-436

Neuronal P-bodies and Transport of microRNA-Repressed mRNAs

Florence Rage

Abstract Highly polarized cells like neurons use specialized RNA transport systems to allow for local control of RNA translation, which is a key to neuronal plasticity in the brain. Several proteins, like ZPB1, FMRP, and Staufen, play an important role in transporting RNA along dendrites to the synapse, and a growing amount of evidence has highlighted the role of miRNA in the control of local RNA translation. P-bodies (Processing bodies) are cytoplasmic structures involved in both RNA degradation and storage of untranslated mRNAs. In neurons, dendritic P-body-like structures (dlP-bodies) are present in the soma and dendrites, sometimes in proximity with synapses. They contain miRNA-repressed mRNA and display motorized movement. Under synaptic activation, dlP-bodies relocalize to more distant sites, exchange molecules with the surrounding cytoplasm, and lose some of their components. We propose a model in which dlP-bodies participate in the transport and local regulation of miRNA targets in the dendrites of mammalian neurons.

1 Introduction

Control of mRNA translation and degradation is an important part of post-transcriptional regulation of gene expression. Cytoplasmic transport of mRNA is a key process linked to local translation, which is required for learning and memory in neurons. Recent studies have shown that neuronal P-bodies might play an important role in these processes.

F. Rage
Institut de Génétique moléculaire de Montpellier, CNRS UMR 5535- IFR 24, 1919 route de Mende, 34293, Montpellier, France

B. de Strooper and Y. Christen (eds.), *Macro Roles for MicroRNAs in the Life and Death of Neurons*, Research and Perspectives in Neurosciences,
DOI 10.1007/978-3-642-04298-0_6,

1.1 MicroRNA: tiny molecules for huge challenges

The last decade has witnessed the discovery of tiny RNAs of 19–25 nucleotides, termed microRNAs (miRNAs), that turned out to form a family of very important regulators of gene expression. The production of miRNA starts in the nucleus with the synthesis of pri-miRNAs, which are then processed into pre-miRNA by Drosha, a RNAse III nuclease, and then exported by Exportin 5. Once in the cytoplasm, pre-miRNAs are recognized and cleaved into mature miRNAs by Dicer, another RNAse III. These mature miRNAs are recruited in a RNA-induced silencing complex (RISC) containing, in particular, the Argonaute (Ago) proteins that directly bind miRNAs and possess an RNAse H fold. The miRNA:Ago complex then recognizes its mRNA target by base pairing, most often generating an imperfect duplex that inhibits mRNA translation (Kim 2005). When the match is perfect, the target mRNA is cleaved by Ago proteins and degraded. More than a thousand mRNAs have been discovered in mammals, and while their function is generally not known, they have already been shown to be involved in diverse biological roles, including temporal patterning during development, cell proliferation and differentiation (Chua et al. 2009; Corbin et al. 2009; Bicker and Schratt 2008).

1.2 Local translation and mRNA transport to synapses

The highly polarized feature of neurons requires that certain processes take place in discrete regions in the cell. In particular, mRNA can be translated not only in the soma but also in growth cones and dendrites. Thus, local translation in growth cones is essential for axonal guidance, whereas in dendrites it represents a fundamental element of neuronal plasticity by allowing an autonomous operation of the post-synapse. Interestingly, expression profiling by microarray analysis revealed that a large set of miRNAs is expressed in the brain (Bak et al. 2008), pointing to a possible role in the control of local translation. Indeed, in 2006, two major studies highlighted the role of microRNAs in synaptic plasticity. In rat, miR-134 was shown to accumulate in dendritic spines and to regulate spine volume (Schratt et al. 2006) by controlling the local translation of limk-1 mRNA. Furthermore, brain-derived neurotrophic factor (BDNF) treatment stimulated limk-1 synthesis by relieving the inhibition mediated by miR-134. In *Drosophila*, stimulation of olfactory neurons resulted in degradation in the synapse of Armitage (a component of RISC) via the proteasome, thus relieving CAMKII mRNA from miR-280 and miR-289 inhibition (Ashraf et al. 2006). Armitage has been shown to be a putative ortholog of mammalian MOV10 protein (Meister et al. 2005).

To reach their destination in the post-synaptic compartment, miRNAs and mRNAs translated locally have to be transported along dendrites. Specific localization generally depends on sequences located in the mRNA 3′ UTR. In many cases, transport occurs within poorly characterized granules that contain RNA-binding proteins and several copies of transported mRNAs. Among the proteins that form granules, the best-characterized are Staufen, cytoplasmic polyadenylation element binding protein (CPEB), zipcode binding protein 1 (ZBP1), survival of motor neuron (SMN), and fragile mental retardation protein (FMRP). All these proteins bind RNA and exhibit motorized movements in neurons (St Johnson 2005.). While they are all present in granules, it is not entirely clear whether they are present in similar or different structures. However, ZBP1 and FMRP showed a high degree of colocalization in hypothalamic neurons, whereas SMN was present in distinct foci (Zhang et al. 2007). Interestingly, FMRP can also interact with miRNA, Dicer and Argonaute (Duan and Jin 2006).

2 P-bodies: properties and function in human cells

2.1 P-bodies are sites of storage for miRNA- repressed mRNAs

P-bodies are discrete cytoplasmic foci that have been recently characterized in many cell types. including yeast and humans. These structures contain many proteins involved in the 5′–3′ mRNA degradation pathway, such as the decapping enzymes Dcp-1 and Dcp-2, the Lsm complex, and the exonuclease Xrn1 (Cougot et al. 2004; Ingelfinger et al. 2002). Furthermore, mRNA decay intermediate can be detected in P-bodies, establishing these structures as mRNA decay centers (Sheth and Parker 2003; Cougot et al. 2004). More recently, proteins of the miRNA pathway have been detected in P-bodies, including GW182, p54/rck, and Ago proteins (Liu et al. 2005; Sen and Blau 2005). The first evidence of a particular localization of miRNAs in the cytoplasm was provided by in situ hybridization studies of let-7 miRNA in human cells, which showed that let-7 probes stained dot-like structures reminiscent of P-bodies (Pillai et al. 2005; Fig. 1A). Indeed, microinjection of in vitro-transcribed, Cy3-labeled, pre-let-7 RNA into the nuclei of HeLa cells revealed that exported let-7 RNA accumulated in or next to P-bodies (Fig. 1B). Furthermore, Renilla Luciferase mRNA reporters containing sequences bound and repressed by let-7 were also transported to P-bodies, whereas reporters containing point mutations that disrupted base pairing with let-7 did not. Moreover, expression of a mutant let-7 miRNA that restored base pairing also restored the localization of the mutant reporter mRNA to P-bodies (Fig. 1C). These results suggested that P-bodies could act not only as degradation centers but also as storage compartments for translationally repressed mRNAs (Pillai et al. 2005), and were further confirmed by the study of CAT-1 mRNA, an amino acid transporter (Bhattacharyya et al. 2006) that is negatively regulated by miR-122 in human hepatoma cells but that can resume translation upon amino-acid starvation. Indeed, while the repressed mRNA accumulated in P-bodies, it was re-directed toward the cytosol upon translational activation.

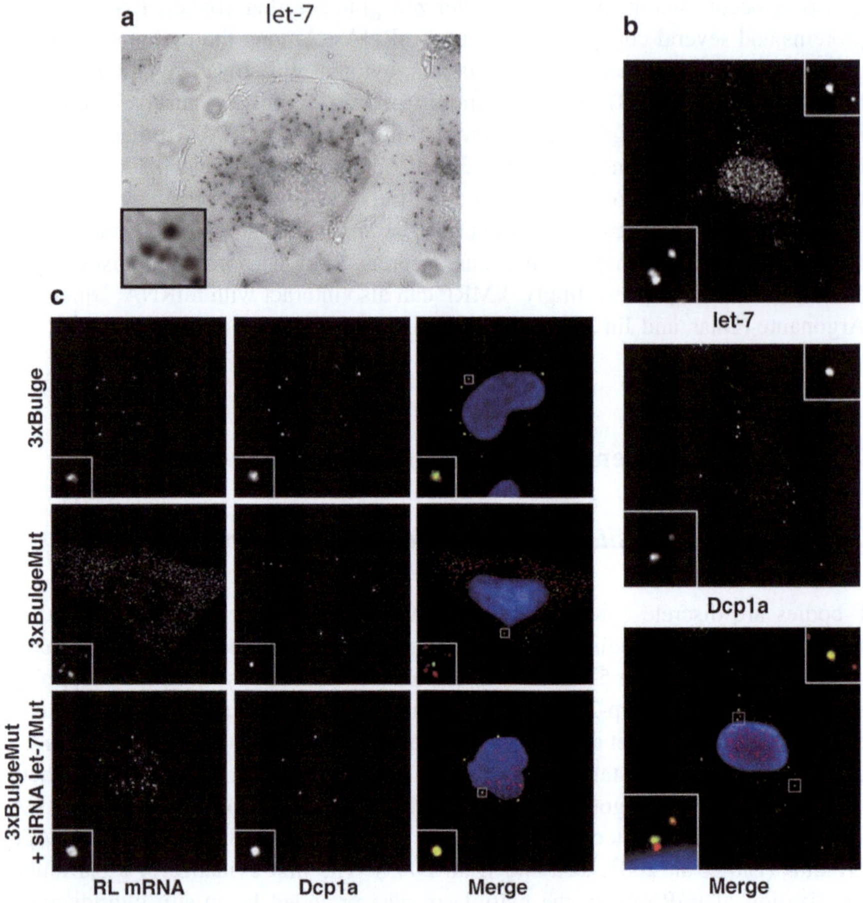

Fig. 1 Localization of translationally repressed mRNA and miRNAs to discrete foci adjacent to or overlapping with P-bodies. Insets represent enlargements of indicated regions, representative of localization of RL reporters, miRNAs, and Dcp1a. (A) Enrichment of endogenous miRNAs let-7 in dot-like structures in HeLa cells. The indicated cells were stained with digoxigenin-labeled complementary probes. (B) Accumulation of let-7 RNA in foci adjacent to or overlapping with P-bodies. Nuclei of HeLa cells were microinjected with an in vitro-transcribed, Cy3-labeled RNA (red), which self-cleaves to produce authentic pre-let-7 RNA. Cells were counterstained with anti-Dcp1a antibody to locate P-bodies (green). (C) HeLa cells were co-transfected with the indicated RL reporters and a plasmid expressing GFP-Dcp1a. Cells shown in a lower panel were co-transfected with the let-7Mut duplex. RL mRNAs were detected by in situ hybridization with Cy3-conjugated probes (red), and Dcp1a was visualized by GFP fluorescence (green)

2.2 P-bodies are sites of degradation for aberrant mRNAs

In *Drosophila*, Staufen is involved in the transport and localization of *oskar* RNA to the posterior pole; this process also requires a complex of four proteins: Mago nashi, Tsunagi, eiF4AIII and Barentsz (Van Eeden et al. 2001; Mohr et al. 2001;

Hatchet and Ephrussi 2001). In mammals, these proteins are the core components of the exon junction complex (EJC), which is deposited at each splice junction by the spliceosome (Le Hir et al. 2000). Remarkably, posterior localization of *oskar* mRNA requires the presence of an intron in its pre-mRNA, indicating that human and yeast complex are assembled by a similar pathway. The EJC core serves as a platform to transiently interact with a number of factors that play diverse role during mRNA metabolism. One of the best-characterized functions of the human EJC residue is the degradation of mRNAs bearing a premature translation termination codon (PTC), also referred to as the NMD pathway. Indeed, while mRNAs normally do not have introns after their stop codons, PTC-containing mRNAs are likely to have one or more, and this triggers its recognition as premature. The EJC plays a fundamental role in this process: it is normally removed from mRNAs by the passage of the first ribosome, but mRNAs containing a stop codon located before a splice junction are unable to do so, and the interaction between stopped ribosomes and the downstream EJC triggers mRNA degradation. Remarkably, when PTC-containing RNAs were stabilized at a late step of the NMD pathway, using either a specific drug or an Xrn1 knock-down, they were shown to accumulate in P-bodies. This accumulation was not observed with the wild-type RNAs, indicating a specific link between NMD and P-bodies (Durand et al. 2008).

In neurons, a recent study highlighted a specific role of the NMD pathway in synaptic plasticity (Giorgi et al. 2007). Indeed, it was shown that neurons expressed an atypical class of mRNAs, which contained introns after their stop codons and thus behaved as constitutive NMD substrates. Remarkably, one of these mRNAs, Arc, is known to be transported in a repressed state to post-synaptic sites in dendrites and to be translated there locally upon synaptic activation. The Arc protein plays a key role in consolidating long-term memory, and Giorgi et al. showed that the presence of an EJC downstream of the stop codon allows for transient pulses of local translation: the mRNA is in a stable form before translation starts, and can be stored in a repressed form, but it is rapidly degraded once translation initiates. Furthermore, the importance of this unique regulation was demonstrated by the alteration of synaptic plasticity following disruption of the NMD pathway.

3 P-bodies form a distinct class of dendritic granules

3.1 Neuronal P-bodies are present in dendrites and are involved in the miRNA pathway

Since NMD and miRNA pathways play important roles in synaptic plasticity and involve P-bodies, it appeared important to characterize these structures in neurons. Using antibodies against Dcp1-a, a canonical component of P-bodies, we and others showed recently that foci reminiscent of P-bodies are present in both the cell body and dendrites of hypothalamic and hippocampal neurons (Fig. 2A). In contrast, no

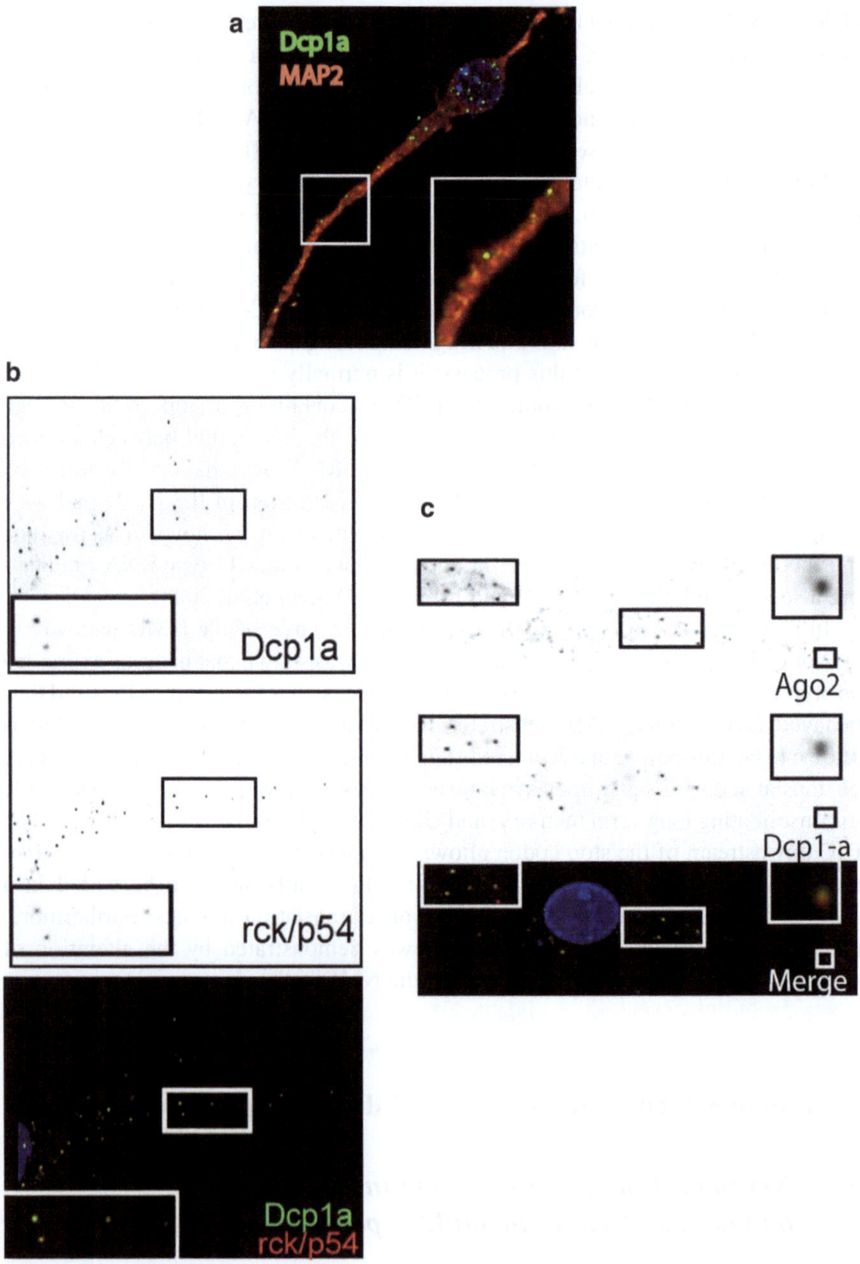

Fig. 2 Dendritic localization of P-body-like structures in rat hippocampal and hypothalamic neurons. In vitro cultured rat hypothalamic neurons were used for immunofluorescence analysis to study the distribution of several P-body markers. (A) Neurons were co-stained for neuronal dendritic marker MAP2. Dcp1a is in green and theMAP2 staining is in red. Nuclei are stained with DAPI (in blue). (B) Foci that were stained with anti-Dcp1a serum were also labeled with anti-RCK/p54 antibodies in processes of neurons. Overlay Dcp1a is in green and rck/p54 is in red. (C) Co-staining of neurons with anti-Dcp1a and anti-Ago2. Overlay: Dcp1a staining in green, Ago2 in red. The insets show enlargements of indicated regions (4x)

foci were identified in axons (Cougot et al. 2008; Zeitelhofer al. 2008). These foci, dubbed dlP-bodies (dendritic-like P-bodies), were sometimes observed in the vicinity of dendritic spines, and purification of synaptoneurosomes indeed confirmed that P-body components were enriched in synapses. These findings mirrored a previous study in Drosophila that showed that P-bodies contain FMRP in dendritic dots (Barbee et al. 2006).

Since many proteins involved in mRNA metabolism accumulate in granule and foci when expressed in neurons, the relationship of these granules and dlP-bodies was investigated by co-localization studies. A large fraction of dlP-bodies was found to coincide with ZBP1 and FMRP granules, whereas SMN and Staufen foci were mostly distinct from dlP-bodies (Cougot et al. 2008; Zeitelhofer al. 2008). In *Drosophila*, neuronal Staufen granules positive for dcp-bodies are very dynamic and allow rapid exchanges with the cytoplasm (Barbee et al. 2006). This finding indicated that some previously characterized neuronal granules corresponded to dlP-bodies, and the partial co-localization observed for many investigated proteins pointed towards heterogeneity of neuronal P-bodies and other granules. Interestingly, dlP-bodies appeared to also be involved in the miRNA pathway: Ago and GW182 co-localized with dlP-bodies labelled with Dcp1-a. Furthermore, in situ hybridization studies showed that several miRNA accumulated in dlP-bodies and that reporter RNAs repressed by miRNA were also present there, at least in the proximal segment of dendrites.

3.2 dlP-bodies are transported along dendrites by motors

Many neuronal RNP granules display motorized movement to reach distant dendritic sites, and dlP-bodies appear to use a similar mechanism. Indeed, time-lapse microscopy experiments showed that dlP-bodies moved bi-directionally along rectilinear trajectories. Detailed analysis of directed movements revealed that, for GFP-Dcp1a, 9.7% were retrograde movements, 14.9% anterograde movements, and 2.3% bi-directional. Furthermore, granules moved an average distance of 5.4 μm (5.5 μm for anterograde movement, 5.3 μm for retrograde movement) and at an average speed of 0.87 μm/sec (0.75 μm/sec for anterograde, 0.99 μm/sec for retrograde). These data are in agreement with the measurements made for ZBP1 (Tiruchinapalli et al. 2003) and for dFMRP, which has been shown to move bidirectionally with a comparable velocity in *Drosophila* dendrites (Ling et al. 2004) and hippocampal neurons (Zeitelhofer et al. 2008). These observations argue that movements of P-bodies in dendrites depend on active transport mediated by molecular motors. Incubation of neurons with nocodazole affects motility but not P-body numbers in dendrites, indicating that they move along microtubules (Zeitelhofer et al. 2008). Interestingly, we found that SMN, which shows little or no colocalization with dlP-bodies, moves over longer distances with a higher velocity. This finding could be explained by the fact that members of the dynein and kinesin superfamily of motor proteins show a wide range of velocities

(Hirokawa et al. 1998). It has previously been shown that the mobility of dendritic ZBP1 granules co-localizing with β-actin mRNA or ZBP1 alone changes in response to KCl-induced depolarization or NMDA receptor blockage (Tiruchinapalli et al. 2003). In agreement, treatment of cultured rat neurons with the glutamate receptor agonist NMDA re-localized dlP-bodies to more distant sites in dendrites. However, it is presently unclear whether the increased transport is specifically oriented towards activated synaptic structures or just represents a general enhancement of granule trafficking. Interestingly, by using different synaptic activators, we observed that dlP-bodies also responded to BDNF and KCl, indicating that they respond to a number of different synaptic stimulations known to be involved in long-term potentiation.

3.3 dlP-bodies become dynamic upon synaptic stimulation

In neurons, mRNPs are stored at synapses and translated locally following activation. Analyzing the dynamic of its components can test the involvement of dlP-bodies in this process. FRAP (Fluorescence recovery after photobleaching) experiments have shown that Dcp1a is rapidly exchanged from P-bodies of non-neuronal cells (Leung et al. 2006). In contrast, it was found that its turnover was very low in dlP-bodies (immobile fraction of 90% after 15 min), indicating that they are stable structures. Remarkably, Dcp1-a exchanged from dlP-bodies much faster following neuronal activation, suggesting that they could release some of their components to the surrounding cytosol. (Fig. 3A, 3B). In support to this possibility, the composition of neuronal P-bodies changes after synaptic stimulation. In hypothalamic neurons, many dlP-bodies lose Ago2 (Fig. 3C, 3D) whereas in hippocampal neurons, they appear to disassemble completely (Zeitelhofer et al. 2008). In hippocampus, the numbers and pattern of dendritic P-bodies remain unchanged when neurons are exposed to oxidative stress (H2O2), showing that the disassembly occurs specifically after synaptic activation and is not due to an artefact of cellular stress (Zeitelhofer et al. 2008).

These results, taken together with other properties of neuronal P-bodies discussed above, make them excellent candidates for structures participating in transport and local regulation of miRNA targets in dendrites of mammalian neurons, as proposed in the model below (Fig. 4).

4 Conclusion and perspectives

The cis-acting sequences important for the localization of dendritic mRNAs are commonly found in their 3'UTRs (Martin and Zukin 2006). Translational repression induced by miRNAs appears sufficient to ensure dendritic localization of a reporter mRNA in the proximal dendritic segment, and one key question is the

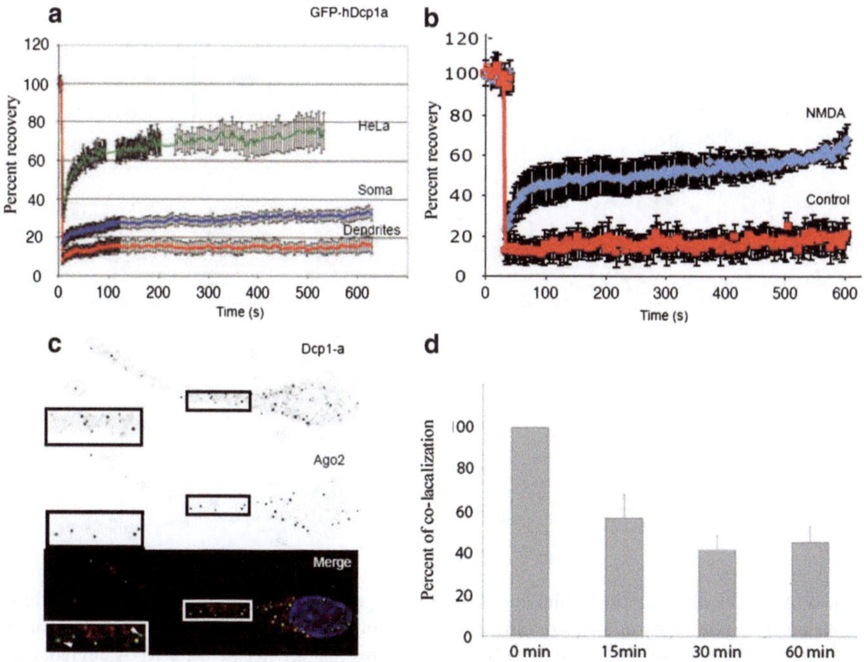

Fig. 3 Synaptic stimulation modifies the exchange rates of dlP-body components and modifies their composition. Hypothalamic neurons and Hela cells expressing GFP-hDcp1a (A) were analyzed by FRAP. Specific foci were bleached after 30-sec recording, and fluorescent signals were recorded over time. The graphs show recovery curves against time for GFPhDcp1a foci. (B) FRAP analysis of GFP-hDcp1 in dendrites before (blue) or after stimulation with 30 μM NMDA (red). (C) Hypothalamic neurons were stimulated (30 μM NMDA for 15 minutes) and analyzed by immunofluorescence with antibodies against Ago2 and hDcp1a. (D) Quantification of the fraction of Dcp1a foci that contained Ago2

nature of the signals that direct the mRNAs to dlP-bodies localized at more distant sites. Similarly, it is currently unclear whether mRNAs ultimately addressed to dlP-bodies are first transported alone to distant sites in dendrites and then to nearby dlP-bodies or if they associate with dlP-bodies in the cell body and are then transported together along dendrites. In any case, one interesting possibility would be that the EJC is involved in the transport of mRNAs to distant dendritic sites. Indeed, in *Drosophila*, *oskar* mRNA is translationally silent and the EJC is required to transport the mRNAs to the posterior pole of *Drosophila* embryos. Although it is expected that, in the absence of other signals, mRNAs constitutively bound by the EJC are rapidly degraded in the cell body, it would be interesting to test whether mRNAs that are bound by the EJC but translationally silent follow a different fate and are transported to distant dendritic sites. Possibly, miRNAs and other translational regulators could synergize with the EJC in this transport process.

Another important question is how mRNAs stored in dlP-bodies in dendrites are reactivated for translation. CPEB protein or neuron-specific ELAV (embryonic lethal abnormal vision) proteins are interesting candidates for factors functioning

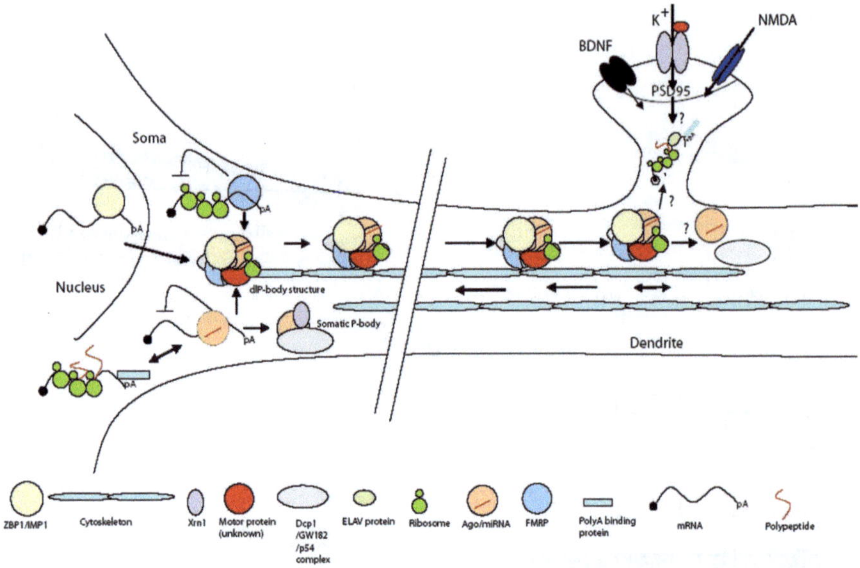

Fig. 4 A model of dendritic trafficking of miRNA-repressed mRNAs in mammalian neurons. Translation repression by miRNPs or other repressor protein factors, i.e., FMRP or ZBP1, ensures entry of the repressed mRNAs into P-body-like structures that are negative for RNA degradation components and positive for ribosomes. These structures are subsequently trafficked to the distal end of the dendrites. These structures are responsive to synaptic stimuli. Local de-repression may occur, allowing translation of the mRNA contained in these dlP-bodies, involving disassembly of the complexes at synapses

in the activation process. HuR, a constitutively expressed ELAV protein, has been documented to relieve mRNA from the miRNA-mediated repression and to promote mRNA exit from P-bodies in non-neuronal cells (Bhattacharyya et al. 2006). It will be interesting to establish whether the neuron-specific ELAV proteins, HuB, HuC and HuD, have similar properties. It is also possible that the RISC complex is modified following activation of the synapse. For instance, stimulation of *Drosophila* neurons promotes the proteasomal degradation of Armitage and allows translation of CAMKII (Ashraf et al. 2006).

References

Ashraf SI, McLoon AL, Sclarsic SM, Kunes S (2006) Synaptic protein synthesis associated with memory is regulated by the RISC pathway in Drosophila. Cell 124:191–205

Bak M, Silahtaroglu A, Møller M, Christensen M, Rath MF, Skryabin B, Tommerup N, Kauppinen S (2008) MicroRNA expression in the adult mouse central nervous system. RNA 14:432–444

Barbee SA, Estes PS, Cziko AM, Hillebrand J, Luedeman RA, Coller JM, Johnson N, Howlett IC, Geng C, Ueda R, Brand AH, Newbury SF, Wilhelm JE, Levine RB, Nakamura A, Parker R, Ramaswami M (2006) Staufen- and FMRP-containing neuronal RNPs are structurally and functionally related to somatic P bodies. Neuron 52:997–1009

Bhattacharyya SN, Habermacher R, Martine U, Closs EI, Filipowicz W (2006) Relief of microRNA-mediated translational repression in human cells subjected to stress. Cell 125:1111–1124
Bicker S, Schratt G (2008) microRNAs: tiny regulators of synapse function in development and disease. J Cell Mol Med 12:1466–1476
Chua JH, Armugam A, Jeyaseelan K (2009) MicroRNAs: Biogenesis, function and applications. Curr Opin Mol Ther 2:189–199
Clark I, Giniger E, Ruohola-Baker H, Jan LY, Jan YN. Transient posterior localization of a kinesin fusion protein reflects anteroposterior polarity of the Drosophila oocyte. Curr Biol 1994:289–300
Corbin R, Olsson-Carter K, Slack F (2009) The role of microRNAs in synaptic development and function. BMB Rept 42:131–135
Cougot N, Babajko S, Seraphin B (2004) Cytoplasmic foci are sites of mRNA decay in human cells. J Cell Biol 165:31–40
Cougot N, Bhattacharyya SN, Tapia-Arancibia L, Bordonné R, Filipowicz W, Bertrand E, Rage F (2008) Dendrites of mammalian neurons contain specialized P-body-like structures that respond to neuronal activation. J Neurosci 28:13793–13804
Duan R, Jin P (2006) Identification of messenger RNAs and microRNAs associated with fragile X mental retardation protein. Methods Mol Biol 342:267–276
Durand S, Cougot N, Mahuteau-Betzer F, Nguyen CH, Grierson DS, Bertrand E, Tazi J, Lejeune F (2007) Inhibition of nonsense-mediated mRNA decay (NMD) by a new chemical molecule reveals the dynamic of NMD factors in P-bodies. J Cell Biol 178:1145–1160
Giorgi C, Yeo GW, Stone ME, Katz DB, Burge C, Turrigiano G, Moore MJ (2007) The EJC factor eIF4AIII modulates synaptic strength and neuronal protein expression.Cell 130:179–191
Hachet O, Ephrussi A (2001) Drosophila Y14 shuttles to the posterior of the oocyte and is required for oskar mRNA transport. Curr Biol 11:1666–1674
Hirokawa N, Noda Y, Okada Y (1998) Kinesin and dynein superfamily proteins in organelle transport and cell division. Curr Opin Cell Biol 10:60–73
Ingelfinger D, Arndt-Jovin DJ, Lührmann R, Achsel T (2002) The human LSm1-7 proteins colocalize with the mRNA-degrading enzymes Dcp1/2 and Xrnl in distinct cytoplasmic foci. RNA 8:1489–1501
Jin P, Alisch RS, Warren ST (2004) RNA and microRNAs in fragile X mental retardation. Nature Cell Biol:6:1048–1053
Kim VN (2005) MicroRNA biogenesis: coordinated cropping and dicing. Nature Rev Mol Cell Biol 6:376–385
Le Hir H, Izaurralde E, Maquat LE, Moore MJ (2000) The spliceosome deposits multiple proteins 20–24 nucleotides upstream of mRNA exon-exon junctions. EMBO J 19:6860–6869
Leung AK, Calabrese JM, Sharp PA (2006) Quantitative analysis of Argonaute protein reveals microRNA-dependent localization to stress granules. Proc Natl Acad Sci USA 103:18125–18130
Lewis BP, Shih IH, Jones-Rhoades MW, Bartel DP, Burge CB (2003) Prediction of mammalian microRNA targets. Cell 115:787–798
Ling SC, Fahrner PS, Greenough WT, Gelfand VI (2004) Transport of Drosophila fragile X mental retardation protein-containing ribonucleoprotein granules by kinesin-1 and cytoplasmic dynein. Proc Natl Acad Sci USA 101:17428–17433
Liu J, Rivas FV, Wohlschlegel J, Yates JR 3rd, Parker R, Hannon GJ (2005) A role for the P-body component GW182 in microRNA function. Nature Cell Biol. 7:1261–1266
Martin KC, Zukin RS (2006) RNA trafficking and local protein synthesis in dendrites: an overview. J Neurosci 26:7131–7134
Meister G, Landthaler M, Peters L, Chen PY, Urlaub H, Lührmann R, Tuschl T (2005) Identification of novel argonaute-associated proteins. Curr Biol 15:2149–2155
Mohr SE Dillon ST, Boswell RE (2001) The RNA-binding protein Tsunagi interacts with Mago Nashi to establish polarity and localize oskar mRNA during Drosophila oogenesis. Genes Dev 15:2886–2899

Pillai RS, Bhattacharyya SN, Artus CG, Zoller T, Cougot N, Basyuk E, Bertrand E, Filipowicz W (2005) Inhibition of translational initiation by Let-7 MicroRNA in human cells. Science 309:1573–1576

Schratt GM, Tuebing F, Nigh EA, Kane CG, Sabatini ME, Kiebler M, Greenberg ME (2006) A brain-specific microRNA regulates dendritic spine development. Nature 439:283–289

Sen GL, Blau HM (2005) Argonaute 2/RISC resides in sites of mammalian mRNA decay known as cytoplasmic bodies. Nature Cell Biol 7:633–636

Sheth U, Parker R (2003) Decapping and decay of messenger RNA occur in cytoplasmic processing bodies. Science 300:805–808

St Johnston D (2005) Moving messages: the intracellular localization of mRNAs. Nature Rev Mol Cell Biol 6:363–375

Van Eeden FJ Palacios IM, Petronczki M, Weston MJ, St Johnston D (2001) Barentsz is essential for the posterior localization of oskar mRNA and colocalizes with it to the posterior pole. J Cell Biol 154:511–523

Tiruchinapalli DM, Oleynikov Y, Kelic S, Shenoy SM, Hartley A, Stanton PK, Singer RH, Bassell GJ (2003) Activity-dependent trafficking and dynamic localization of zipcode binding protein 1 and beta-actin mRNA in dendrites and spines of hippocampal neurons. J Neurosci 23:3251–3261

Zeitelhofer M, Karra D, Macchi P, Tolino M, Thomas S, Schwarz M, Kiebler M, Dahm R (2008) Dynamic interaction between P-bodies and transport ribonucleoprotein particles in dendrites of mature hippocampal neurons. J Neurosci 28:7555–7562

Zhang H, Xing L, Singer RH, Bassell GJ (2007) QNQKE targeting motif for the SMN-Gemin multiprotein complexin neurons. J Neurosci Res 85:2657–2667

Crosstalk between microRNA and Epigenetic Regulation in Stem Cells

Keith Szulwach, Shuang Chang, and Peng Jin

Abstract The differential expression of common genomes in many multicellular organisms is the fundamental phenomenon studied in epigenetics. Stem cells retain the unique ability to hold potency and self renew as well as the ability to differentiate into distinct cell types upon receipt of specific environmental cues. These processes exemplify the critical role of epigenetic regulation in modulating the expression of common genomes and provide a link between cellular genotype and phenotype. Small regulatory RNAs, 21 to 30 nucleotides in length, including microRNAs (miRNAs), are sequence-specific posttranscriptional regulators of thousands of target messenger RNAs (mRNAs) and therefore shape diverse cellular pathways. Multiple forms of epigenetic regulation within the context of stem cells, specifically that of epigenetic regulation of microRNAs, have been found to hold significance during neurogenesis from a stem cell state. Mechanisms by which this regulation is accomplished and discussion of the role of epigenetic regulation of miRNA expression during stem cell function will be put forth.

1 Introduction

Epigenetic regulation is defined as heritable changes in the function of genetic elements without changes in the actual genetic or underlying DNA sequence. The epigenetic regulation of genomes in mammalian systems has allowed for the unique ability of particular cells to not only produce differential progeny but to maintain the ability to self renew, continually retaining the potency to produce unique cell types in subsequent cell divisions. Cells having these characteristics are formally known as stem cells.

P. Jin (✉)
Department of Human Genetics and Graduate Program in Genetics and Molecular Biology, Emory University School of Medicine, 615 Michael Street, Room 323, Atlanta, GA30322, USA
e-mail: peng.jin@emory.edu

B. de Strooper and Y. Christen (eds.), *Macro Roles for MicroRNAs in the Life and Death of Neurons*, Research and Perspectives in Neurosciences,
DOI 10.1007/978-3-642-04298-0_7,

In mammalian systems, the molecular mechanisms contributing to stem cell function and epigenetic regulation of genomic information are quite varied and are still not completely understood. Such mechanisms include, but may not necessarily be limited to, those directly influencing DNA accessibility and gene expression in the context of chromatin, such as that of covalent and non-covalent chemical modifications to both DNA and histones. Other mechanisms, including those indirectly influencing the information flow from DNA to protein, like alternative splicing and poly-adenylation of mRNA transcripts, post-translational modifications of proteins, and posttranscriptional regulation of transcribed RNA, may also be considered epigenetic under strict definitions. Perhaps one of the most intriguing of the mechanisms that have been described as influencing epigenetic processes is that involving some non-coding RNA transcripts. Reports of epigenetic influence by non-coding RNA, that is itself encoded in the genome, offers the possibility that these RNAs may be both subjected to and directing epigenetic regulation (Plath et al. 2003; Sanchez-Elsner et al. 2006; Rinn et al. 2007). The obvious implication of such observations is the potential for such non-coding RNAs as ideal molecular links between cellular genotype and phenotype. Indeed, non-coding RNAs are known to be involved in direct and indirect epigenetic regulation of genetic information in the context of stem cells as well. As a result, we have hypothesized that the epigenetic regulation of non-coding RNA, as well as the epigenetic influence of that non-coding RNA on cellular phenotype, contribute to stem cell function. Here, we will discuss various epigenetic mechanisms related to stem cell function, emphasizing the potential for epigenetic regulation of non-coding RNA, particularly microRNA (miRNA). We will provide a framework for how epigenetic regulation of and by miRNAs may influence the function of stem cells and differentiation of stem cells toward neuronal lineages.

2 Adult Neural Stem Cells and Neurogenesis

In adults, stem cells exist in many tissues throughout life and may play critical roles in tissue regeneration and repair. Neural stem cells (NSCs) are multipotent cells that are characterized by their abilities to self-renew and to generate differentiated cells, specifically in the central nervous system. Neurogenesis is defined as the process of generating new neurons from NSCs, which consists of the proliferation and fate determination of NSCs, migration and survival of young neurons, and maturation and integration of newly matured neurons (Ming and Song 2005). Since the discovery of adult neurogenesis, neuroscientists and developmental biologists have been exploring the regulatory mechanisms and functions of this fascinating process. Our current knowledge supports a model whereby adult neurogenesis is regulated by both intrinsic programs and extrinsic modulators. Intrinsic programs include genes, genetic background, and epigenetic modifications that are essential for controlling NSC self-renewal and multipotency. Extrinsic factors include both the microenvironment in which NSCs physically reside and the stimuli that NSCs receive due to endocrinal, physiological and pathological changes (Zhao et al. 2004). Although many significant advances have been made, more challenges are ahead of us in understanding how these regulatory mechanisms coordinately modulate neurogenesis and define a neurogenic niche in adult mammalian brains. In particular, the potential epigenetic regulation of miRNA-driven regulatory pathways may be crucial in understanding the molecular mechanisms contributing to NSC epigenesis and the production of new neurons throughout adulthood.

3 Epigenetic Regulation in Stem Cells

3.1 DNA methylation

DNA methylation is a covalent modification of cytosine at the position C-5 in CpG dinucleotides. In mammals, over 70% of CpG dinucleotides are methylated and nearly all DNA methylation occurs on CpG dinucleotides. Concentrations of unmethylated CpG dinucleotides are usually found in the promoters and the first exons of active protein coding genes, termed CpG islands (Jones and Takai 2001). Conversely, methylated CpG islands are generally associated with a condensed chromatin state that is repressive toward transcription of associated DNA. As a consequence, differential states of DNA methylation may modulate the expression of underlying genetic information and so are considered epigenetic in nature. DNA methylation is catalyzed by three DNA methyltransferase proteins (DNMTs). The de novo establishment of DNA methylation relies on DNMT3a and DNMT3b, whereas the maintenance of DNA methylation depends on DNMT1, which specifically recognizes hemi-methylated DNA and methylates the remaining strand (Jaenisch and Bird 2003).

Mammalian DNA methylation has been implicated in a diverse range of cellular functions, including tissue-specific gene expression, cell differentiation, genomic imprinting, and X chromosome inactivation (Bird 2002). DNA methylation represses gene expression by either directly blocking the binding of transcription factors (Takizawa et al. 2001) or recruiting a family of methylated-CpG binding proteins (MBDs), many of which share homology only in their methyl-CpG binding (MBD) domains (Bird 2002). Therefore, DNA methylation is thought to be a critical factor in regulating the expression of common genomes in certain cellular or environmental contexts and so also in regulating the potency of stem cells.

3.2 MBD protein family

The MBD protein family includes MBD1, MBD2, MBD3, MBD4, MeCP2, Keiso, and several newly discovered members (Klose and Bird 2006). MBD1/Mbd1 is a multifunctional protein that is localized to both euchromatin and heterochromatin, with influence in NSC function. MBD1 has two DNA-binding domains that specifically recognize methylated CpGs and a zinc finger (CXXC3) domain that specifically binds unmethylated CpGs. The presence of two DNA binding domains in MBD1 may contribute to higher affinity and specificity in binding DNA sequences (Jorgensen et al. 2004). Transcriptional repression by MBD1 can be facilitated by several putative cofactors (Fujita et al. 2003a), and it is likely that MBD1 represses transcription through various mechanisms, depending on the particular gene and cell type. However, despite great effort, few MBD1 target genes have been identified. Moreover, although recent literature suggests that each MBD protein may have its own preferred binding sites in the genome (Klose et al. 2005), currently available structure-function data have not provided sequence specificity other than CpGs. Extensive in vitro analyses have suggested a role for MBD1 in transcriptional repression (Fujita et al. 2003b), chromatin assembly (Fujita et al. 2003b; Sarraf and Stancheva 2004), and heterochromatin structure maintenance; however, the biological function of MBD1 remains relatively unknown (Setoguchi et al. 2006).

3.3 Mbd1 may regulate adult neural stem/progenitor cells

Despite the relative lack of data on its role in chromatin function, Mbd1 has been found to be localized in both neurons and a subset of Nestin-positive immature cells in the germinal zone of the hippocampus (SGZ) of adult mice (Zhao et al. 2003). This finding suggests that Mbd1 may regulate functions of adult neural stem/ progenitor cells (NSPCs) via modulating epigenetic control of gene expression. Mbd1 mutant (*Mbd1*$^{-/-}$) mice develop normally into adulthood, with no detectable developmental defects except for mild reduction in forebrain weight, indicating that early development of the brain may be suboptimal in the absence of Mbd1. In the *Mbd1*$^{-/-}$ hippocampus, cell proliferation is normal, but the survival of newborn cells is significantly reduced along with decreased neuronal differentiation capacity of NSPCs. As a possible consequence, the dentate gyrus in *Mbd1*$^{-/-}$ mice has reduced cell density. In addition, adult *Mbd1*$^{-/-}$ mice have spatial learning deficits and markedly reduced dentate gyrus-specific long term potentiation, a proposed cellular mechanism for learning and memory (Zhao et al. 2003). At the cellular level, NSPCs isolated from adult *Mbd1*$^{-/-}$ mice have a reduced neuronal differentiation capacity in vitro, consistent with our in vivo findings (Zhao et al. 2003). In addition, *Mbd1*$^{-/-}$ NSPCs have increased genomic instability and increased expression of endogenous stem cell mitogen fibroblast growth factor (Fgf-2). Fgf-2 is a potent growth factor for a large number of cell types, and its overexpression has been found in many transformed tumor cells, including glioma cells with possible NSC origin (Ueba et al. 1999). These studies of the function of Mbd1 in NSPCs provide a key example of the influence epigenetic modulation may have on stem cell function and neurogenesis.

3.4 *Histone modifications regulate adult NSCs and neurogenesis*

In eukaryotic cells, the basic unit of chromatin is formed by 146 base pairs of DNA wrapped around the histone octamer. The core histones, H2A, H2B, H3, and H4, are subject to numerous and varied modifications, including acetylation, methylation, and phosphorylation. Among these modifications, lysine (K) acetylation and methylation are the best understood (Bernstein et al. 2007). Initial histone modification studies focused largely on acetylation, which is catalyzed by two opposing enzymes, histone acetyltransferease (HAT) and histone deacetylase (HDAC). At least eight HATs and nine HDACs have been identified in mammals (Miremadi et al. 2007). The activities of HATs and HDACs can directly affect adult NSCs and adult neurogenesis. For example, neuron-specific genes share the conserved 21–23-basepair DNA response element, RE-1 (repressor element 1). Neuronal restricted silencing factor (NRSF or REST) binds to RE-1 and forms a repressing complex that represses neuronal gene expression in non-neuronal cells by recruiting HDAC1/2 and Sin3A (Stanfield and Trice 1988; Lunyak et al. 2002; Ballas et al. 2005; Lunyak and Rosenfeld 2005). Treatment of adult NSPCs by volporic acid (VPA), a HDAC inhibitor and antiepileptic medicine, leads to reduced proliferation, increased neuronal differentiation, and decreased astrocyte and oligodendrocyte differentiation through activating a pan-neuronal transcription factor NeuroD1 (Kuwabara et al. 2004). More recently, Jessberger et al. (2007) further confirmed that VPA treatment attenuates seizure-induced aberrant neurogenesis through regulating NRSF and HDACs. In the developing brain, VPA administration also induces significant hypomyelination and delay in the differentiation of oligodendrocytes through inhibiting the activity of HDACs (Shen et al. 2005). Histone methylation plays important roles in embryonic stem cell (ESC) development, cell fate determination, and X chromosome inactivation (Plath et al. 2003; Torres-Padilla et al. 2007). Patterns of histone H3K4 methylation (an active chromatin mark), H3K27 methylation (a temporarily inactive chromatin mark), and H3K9 methylation (a long-term repressive chromatin mark) define the chromatin state of NSCs (Mikkelsen et al. 2007). Mutation of Bmi-I, a component of polycomb group proteins with H3K27 methylase activity, results in reduced self-renewal of adult NSCs (Molofsky et al. 2005). The chromatin state of NSCs is distinct from that of ESCs or differentiated cell types (Mikkelsen et al. 2007).

It is likely that, during NSC differentiation, the chromatin state that defines the NSC signature shifts towards a state corresponding to that of more differentiated cell types. Therefore, the temporal epigenetic landscape of NSCs could in fact be a much more precise marker for stem cell signature than the expression patterns of single genes.

4 MiRNA function in NSCs and Neurogenesis

MiRNAs were originally discovered in genetic screens identifying heterochronic developmental regulators in *C. elegans* (Lee et al. 1993; Wightman et al. 1993). More recent observations continue to support roles for miRNAs in determining and maintaining developmental cell fates both spatially and temporally. In particular, miRNAs are known to act in a cell type-specific manner and, furthermore, have been found to be expressed at especially high levels in the central nervous system, acting critically during neurogenesis and neuronal patterning (Lagos-Quintana et al. 2002; Babak et al. 2004; Barad et al. 2004; Kim et al. 2004; Miska et al. 2004; Sempere et al. 2004). It has also been shown that the expression patterns of small RNA undergo dynamic changes during the differentiation of human ESCs into neuronal progenitors and mature neurons (Landgraf et al. 2007; Wu et al. 2007).

One of the most well-characterized examples of a miRNA having influence on neural development from the stem cell state is that of mir-124a, a miRNA with preferentially high expression in brain. Exogenous expression of the neuron-enriched mir-124a, as well as another highly neuron-enriched miRNA, mir-9, in ESCs has been found to inhibit their differentiation into astrocytes by modulating the STAT3 pathway, which is critical for astrocyte differentiation (Kim et al. 2004). In addition, mir-124a has been found to down-regulate the expression of the small C-terminal domain phosphatase 1 (an anti-neuronal phosphatase) and promote pro-neuronal RNA splicing and thus neurogenesis (Makeyev et al. 2007; Visvanathan et al. 2007).

In fact, there is also precedence for the involvement of epigenetic regulation of mir-124a in the context of neurogenesis from the stem cell state, further indicating the potential for epigenetic regulation of miRNA being critical to the proper function of stem cells and neurogenesis. mir-124a has been shown to be regulated by a transcriptional regulatory complex that also happens to be associated with an epigenetic effector and DNA methyl CpG binding protein of the same family as MBD1, methyl CpG binding protein 2 (MeCP2), during neurogenesis from P19 embryonic carcinoma cells (Conaco et al. 2006). P19 cells have stem cell character in that they are able to maintain a certain amount of potency in culture and can be induced to differentiate into neuronal cell types with the treatment of retinoic acid. MeCP2 has been shown to recruit the histone lysine methyltransferase, SUV39H1, as part of the co-repressor complex specific for neuronal genes that was previously mentioned, REST or NRSF (Shahbazian and Zoghbi 2002; Moretti and Zoghbi 2006). REST has itself been shown to directly bind regions proximal to a family of miRNAs including mir-9, mir-124a, and mir-132. Reduced expression of REST during retinoic acid-induced neuronal differentiation of P19 cells was shown to correlate with increased expression of mir-124a and decreased expression of non-neuronal mRNAs, thus revealing the potential for reciprocal epigenetic action of REST and mir-124a in P19-derived neurogenesis (Conaco et al. 2006). This example, similar to that of MBD1, again emphasizes a critical role for epigenetic regulation in NSC function and further implicates the potential for epigenetic regulation of miRNA expression in such processes.

Beyond mir-124a, there is also intriguing evidence for critical epigenetic regulation of miRNA in the context of stem cells. This evidence comes from genome-wide scans of binding sites for key transcriptional and epigenetic regulators in ESCs. These studies have identified core transcriptional regulatory networks involving the factors Oct4, Sox2, and Nanog as well as binding sites for the

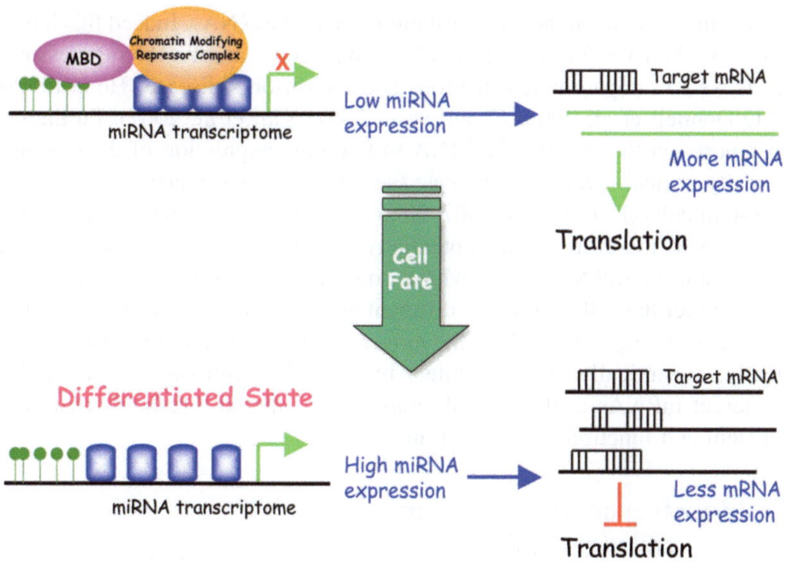

Fig. 1 Generalization of a model involving epigenetic regulation of miRNAs in determining cell fate. A miRNA may be epigenetically silenced in the undifferentiated cell state by the combination of a methyl-DNA binding protein and associated chromatin-modifying repressor complex. In this state, the expression of the repressed miRNA is low whereas the expression of target mRNA transcript or protein coded for by the target transcript is high relative to the differentiated state. Upon an extrinsic or intrinsic signal cue for cellular differentiation, epigenetic repression of the target miRNA is released, and miRNA expression increases. The increase in miRNA expression then correlates with decreased stability or translation of target mRNAs. By epigenetically regulating one miRNA, the cell can thereby direct and fine-tune expression of multiple miRNA-targeted mRNA transcripts during cell fate determination. The illustrated example provides a single direction in which this mechanism may potentially function, and it may be equally likely that the mechanism contributes to cell fate determination in the opposite manner. In this case, miRNA expression may become epigenetically silenced during differentiation. Subsequently, target mRNA translation would increase and proteins important to the differentiating cell would be expressed at higher levels than in the undifferentiated cell

epigenetic regulatory Polycomb proteins. Sox2, in particular, has been shown to play a critical role in differentiation of NSCs specifically toward the neuronal lineage. In both studies, the above transcriptional and epigenetic regulators were found to directly interact with genomic regions proximal to specific miRNAs in ESCs (Boyer et al., 2005, 2006; Lee et al. 2006). This finding strongly indicates the importance of these miRNA in either the maintenance or differentiation of stem cells.

The implications for epigenetic regulation of miRNA expression in the context of stem cells are potentially far-reaching. Reports on the abilities of individual miRNA to target multiple mRNA transcripts simultaneously and the potential of individual mRNA transcripts to be regulated by multiple different miRNAs simultaneously may in fact allow for the potential of particular miRNAs or subsets of miRNAs to drive regulatory pathways during cellular, multicellular, and organism development via their impact on translation of target mRNAs. Indeed this has been shown to be the case for some miRNAs in oncogenic and tumor suppressor networks as well in dopaminergic neurons and Parkinson's disease (He et al. 2005, 2007; O'Donnell et al. 2005; Chang et al. 2007; Kim et al. 2007). Furthermore, recent reports on the ability of miRNA to fine-tune expression of their targets to biologically critical levels also indicate the potential importance of such mechanisms in stem cells (Karres et al. 2007; Xiao et al. 2007). Epigenetic regulation of a single miRNA, therefore, could potentially set in motion the regulation or fine tuning of multiple mRNA targets whose downstream expression and function are critical to either the self-renewal or differentiation of a stem cell. By regulating the expression of a single miRNA at the epigenetic level, a cell could thereby more efficiently, dynamically, and accurately regulate the expression levels of downstream target mRNAs at the critical temporal and spatial thresholds required for proper stem cell function (see Figs. 1 and 2).

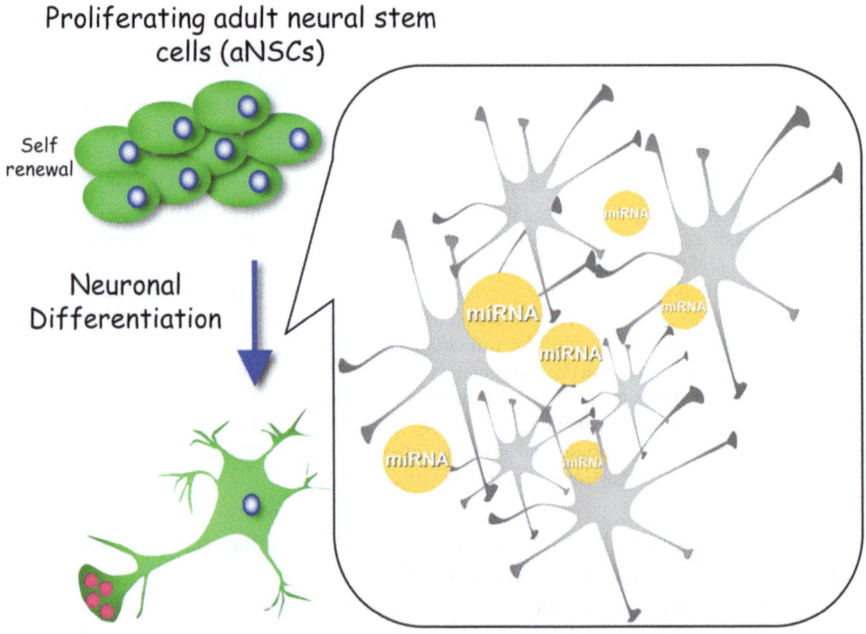

Fig. 2 Potential action of multiple interacting and epigenetically influenced miRNA regulatory circuits involved in neuronal differentiation and/or stem cell function. Epigenetic regulation of various miRNA to different levels of expression, as represented by the different sizes of the "miRNA dots," may work simultaneously to fine-tune the expression levels of multiple mRNA targets, represented by the "mRNA spokes" overlapping the "miRNA dots." Meanwhile, multiple protein products from connected "mRNA spokes" may work together within pathways, as represented by the central core connecting the "spokes," driving stem cell function and/or neuronal differentiation. Also illustrated here is the potential for overlap between these pathways themselves, or overlap of "spokes." Therefore, the proper modulation of miRNA expression at an epigenetic level becomes critical to the proper simultaneous fine-tuning of multiple mRNA targets within interacting pathways driving neuronal differentiation and influencing stem cell function

5 Conclusion

The establishment of cell type-specific systems is imperative toward understanding epigenetic mechanisms contributing to the control and function of stem cells. Identification of key miRNAs involved in the maintenance/self-renewal and differentiation of adult NSPCs could possibly overlay additional molecular and genetic variables, in particular those concerned with epigenetic mechanisms, to further identify epigenetic and miRNA-related regulatory pathways that are critical to proper stem cell function. Identification of such regulatory pathways may in fact prove to be central to stem cell biology, since the epigenetic regulation miRNAs, which themselves can regulate multiple mRNA transcripts, would provide a means by which stem cells might efficiently, dynamically, and accurately fine-tune their own function. Additionally, employing mechanisms of epigenetic control may also provide a means by which stem cells could respond to extrinsic or environmental cues to respond, transfer, and fix information onto the genome so that proper function might be carried out.

Acknowledgments P.J. is supported by International Rett Syndrome Foundation and National Institutes of Health. P.J. is a recipient of the Beckman Young Investigator Award and the Basil O'Connor Scholar Research Award, as well as an Alfred P. Sloan Research Fellow in Neuroscience.

References

Babak T, Zhang W, Morris Q, Blencowe BJ, Hughes TR (2004) Probing microRNAs with microarrays: tissue specificity and functional inference. RNA 10:1813–1819

Ballas N, Grunseich C, Lu DD, Speh JC, Mandel G (2005) REST and its corepressors mediate plasticity of neuronal gene chromatin throughout neurogenesis. Cell 121:645–657

Barad O, Meiri E, Avniel A, Aharonov R, Barzilai A, Bentwich I, Einav U, Gilad S, Hurban P, Karov Y, Lobenhofer EK, Sharon E, Shiboleth YM, Shtutman M, Bentwich Z, Einat P (2004) MicroRNA expression detected by oligonucleotide microarrays: system establishment and expression profiling in human tissues. Genome Res 14:2486–2494

Bernstein BE, Meissner A, Lander ES (2007) The mammalian epigenome. Cell 128:669–681

Bird A (2002) DNA methylation patterns and epigenetic memory. Genes Dev 16:6–21

Boyer LA, Lee TI, Cole MF, Johnstone SE, Levine SS, Zucker JP, Guenther MG, Kumar RM, Murray HL, Jenner RG, Gifford DK, Melton DA, Jaenisch R, Young RA (2005) Core transcriptional regulatory circuitry in human embryonic stem cells. Cell 122:947–956

Boyer LA, Plath K, Zeitlinger J, Brambrink T, Medeiros LA, Lee TI, Levine SS, Wernig M, Tajonar A, Ray MK, Bell GW, Otte AP, Vidal M, Gifford DK, Young RA, Jaenisch R (2006) Polycomb complexes repress developmental regulators in murine embryonic stem cells. Nature 441:349–353

Chang TC, Wentzel EA, Kent OA, Ramachandran K, Mullendore M, Lee KH, Feldmann G, Yamakuchi M, Ferlito M, Lowenstein CJ, Arking DE, Beer MA, Maitra A, Mendell JT (2007) Transactivation of miR-34a by p53 broadly influences gene expression and promotes apoptosis. Mol Cell 26:745–752

Chen C, Ridzon DA, Broomer AJ, Zhou Z, Lee DH, Nguyen JT, Barbisin M, Xu NL, Mahuvakar VR, Andersen MR, Lao KQ, Livak KJ, Guegler KJ (2005) Real-time quantification of microRNAs by stem-loop RT-PCR. Nucleic Acids Res 33:e179

Conaco C, Otto S, Han JJ, Mandel G (2006) Reciprocal actions of REST and a microRNA promote neuronal identity. Proc Natl Acad Sci USA 103:2422–2427

Fujita N, Watanabe S, Ichimura T, Ohkuma Y, Chiba T, Saya H, Nakao M (2003a) MCAF mediates MBD1-dependent transcriptional repression. Mol Cell Biol 23:2834–2843

Fujita N, Watanabe S, Ichimura T, Tsuruzoe S, Shinkai Y, Tachibana M, Chiba T, Nakao M (2003b) Methyl-CpG binding domain 1 (MBD1) interacts with the Suv39h1-HP1 heterochromatic complex for DNA methylation-based transcriptional repression. J Biol Chem 278:24132–24138

He L, Thomson JM, Hemann MT, Hernando-Monge E, Mu D, Goodson S, Powers S, Cordon-Cardo C, Lowe SW, Hannon GJ, Hammond SM (2005) A microRNA polycistron as a potential human oncogene. Nature 435:828–833

He L, He X, Lim LP, de Stanchina E, Xuan Z, Liang Y, Xue W, Zender L, Magnus J, Ridzon D, Jackson AL, Linsley PS, Chen C, Lowe SW, Cleary MA, Hannon GJ (2007) A microRNA component of the p53 tumour suppressor network. Nature 447:1130–1134

Jaenisch R, Bird A (2003) Epigenetic regulation of gene expression: how the genome integrates intrinsic and environmental signals. Nature Genet 33 Suppl:245–254

Jessberger S, Nakashima K, Clemenson GD, Jr., Mejia E, Mathews E, Ure K, Ogawa S, Sinton CM, Gage FH, Hsieh J (2007) Epigenetic modulation of seizure-induced neurogenesis and cognitive decline. J Neurosci 27:5967–5975

Jones PA, Takai D (2001) The role of DNA methylation in mammalian epigenetics. Science 293:1068–1070

Jorgensen HF, Ben-Porath I, Bird AP (2004) Mbd1 is recruited to both methylated and nonmethylated CpGs via distinct DNA binding domains. Mol Cell Biol 24:3387–3395

Karres JS, Hilgers V, Carrera I, Treisman J, Cohen SM (2007) The conserved microRNA miR-8 tunes atrophin levels to prevent neurodegeneration in Drosophila. Cell 131:136–145

Kim J, Krichevsky A, Grad Y, Hayes GD, Kosik KS, Church GM, Ruvkun G (2004) Identification of many microRNAs that copurify with polyribosomes in mammalian neurons. Proc Natl Acad Sci USA 101:360–365

Kim J, Inoue K, Ishii J, Vanti WB, Voronov SV, Murchison E, Hannon G, Abeliovich A (2007) A MicroRNA feedback circuit in midbrain dopamine neurons. Science 317:1220–1224

Klose RJ, Bird AP (2006) Genomic DNA methylation: the mark and its mediators. Trends Biochem Sci 31:89–97

Klose RJ, Sarraf SA, Schmiedeberg L, McDermott SM, Stancheva I, Bird AP (2005) DNA binding selectivity of MeCP2 due to a requirement for A/T sequences adjacent to methyl-CpG. Mol Cell 19:667–678

Kuwabara T, Hsieh J, Nakashima K, Taira K, Gage FH (2004) A small modulatory dsRNA specifies the fate of adult neural stem cells. Cell 116:779–793

Lagos-Quintana M, Rauhut R, Lendeckel W, Tuschl T (2001) Identification of novel genes coding for small expressed RNAs. Science 294:853–858

Lagos-Quintana M, Rauhut R, Yalcin A, Meyer J, Lendeckel W, Tuschl T (2002) Identification of tissue-specific microRNAs from mouse. Curr Biol 12:735–739

Landgraf P, Rusu M, Sheridan R, Sewer A, Iovino N, Aravin A, Pfeffer S, Rice A, Kamphorst AO, Landthaler M, Lin C, Socci ND, Hermida L, Fulci V, Chiaretti S, Foa R, Schliwka J, Fuchs U, Novosel A, Muller RU, Schermer B, Bissels U, Inman J, Phan Q, Chien M, Weir DB, Choksi R, De Vita G, Frezzetti D, Trompeter HI, Hornung V, Teng G, Hartmann G, Palkovits M, Di Lauro R, Wernet P, Macino G, Rogler CE, Nagle JW, Ju J, Papavasiliou FN, Benzing T, Lichter P, Tam W, Brownstein MJ, Bosio A, Borkhardt A, Russo JJ, Sander C, Zavolan M, Tuschl T (2007) A mammalian microRNA expression atlas based on small RNA library sequencing. Cell 129:1401–1414

Lau NC, Lim LP, Weinstein EG, Bartel DP (2001) An abundant class of tiny RNAs with probable regulatory roles in Caenorhabditis elegans. Science 294:858–862

Lee RC, Ambros V (2001) An extensive class of small RNAs in Caenorhabditis elegans. Science 294:862–864

Lee RC, Feinbaum RL, Ambros V (1993) The C. elegans heterochronic gene lin-4 encodes small RNAs with antisense complementarity to lin-14. Cell 75:843–854

Lee TI, Jenner RG, Boyer LA, Guenther MG, Levine SS, Kumar RM, Chevalier B, Johnstone SE, Cole MF, Isono K, Koseki H, Fuchikami T, Abe K, Murray HL, Zucker JP, Yuan B, Bell GW, Herbolsheimer E, Hannett NM, Sun K, Odom DT, Otte AP, Volkert TL, Bartel DP, Melton DA, Gifford DK, Jaenisch R, Young RA (2006) Control of developmental regulators by Polycomb in human embryonic stem cells. Cell 125:301–313

Lunyak VV, Rosenfeld MG (2005) No rest for REST: REST/NRSF regulation of neurogenesis. Cell 121:499–501

Lunyak VV, Burgess R, Prefontaine GG, Nelson C, Sze SH, Chenoweth J, Schwartz P, Pevzner PA, Glass C, Mandel G, Rosenfeld MG (2002) Corepressor-dependent silencing of chromosomal regions encoding neuronal genes. Science 298:1747–1752

Makeyev EV, Zhang J, Carrasco MA, Maniatis T (2007) The MicroRNA miR-124 promotes neuronal differentiation by triggering brain-specific alternative pre-mRNA splicing. Mol Cell 27:435–448

Mikkelsen TS, Ku M, Jaffe DB, Issac B, Lieberman E, Giannoukos G, Alvarez P, Brockman W, Kim TK, Koche RP, Lee W, Mendenhall E, O'Donovan A, Presser A, Russ C, Xie X, Meissner A, Wernig M, Jaenisch R, Nusbaum C, Lander ES, Bernstein BE (2007) Genome-wide maps of chromatin state in pluripotent and lineage-committed cells. Nature 448:553–560

Ming GL, Song H (2005) Adult neurogenesis in the mammalian central nervous system. Annu Rev Neurosci 28:223–250

Miremadi A, Oestergaard MZ, Pharoah PD, Caldas C (2007) Cancer genetics of epigenetic genes. Human Mol Genet 16 Spec No 1:R28–49

Miska EA, Alvarez-Saavedra E, Townsend M, Yoshii A, Sestan N, Rakic P, Constantine-Paton M, Horvitz HR (2004) Microarray analysis of microRNA expression in the developing mammalian brain. Genome Biol 5:R68

Molofsky AV, He S, Bydon M, Morrison SJ, Pardal R (2005) Bmi-1 promotes neural stem cell self-renewal and neural development but not mouse growth and survival by repressing the p16Ink4a and p19Arf senescence pathways. Genes Dev 19:1432–1437

Moretti P, Zoghbi HY (2006) MeCP2 dysfunction in Rett syndrome and related disorders. Curr Opin Genet Dev 16:276–281

Muotri AR, Chu VT, Marchetto MC, Deng W, Moran JV, Gage FH (2005) Somatic mosaicism in neuronal precursor cells mediated by L1 retrotransposition. Nature 435:903–910

O'Donnell KA, Wentzel EA, Zeller KI, Dang CV, Mendell JT (2005) c-Myc-regulated microRNAs modulate E2F1 expression. Nature 435:839–843

Plath K, Fang J, Mlynarczyk-Evans SK, Cao R, Worringer KA, Wang H, de la Cruz CC, Otte AP, Panning B, Zhang Y (2003) Role of histone H3 lysine 27 methylation in X inactivation. Science 300:131–135

Reynolds BA, Weiss S (1996) Clonal and population analyses demonstrate that an EGF-responsive mammalian embryonic CNS precursor is a stem cell. Dev Biol 175:1–13

Reynolds BA, Weiss, S (1992) Generation of neurons and astrocytes from isolated cells of the adult mammalian central nervous system. Science 255:1646

Rinn JL, Kertesz M, Wang JK, Squazzo SL, Xu X, Brugmann SA, Goodnough LH, Helms JA, Farnham PJ, Segal E, Chang HY (2007) Functional demarcation of active and silent chromatin domains in human HOX loci by noncoding RNAs. Cell 129:1311–1323

Sanchez-Elsner T, Gou D, Kremmer E, Sauer F (2006) Noncoding RNAs of trithorax response elements recruit Drosophila Ash1 to Ultrabithorax. Science 311:1118–1123

Sarraf SA, Stancheva I (2004) Methyl-CpG binding protein MBD1 couples histone H3 methylation at lysine 9 by SETDB1 to DNA replication and chromatin assembly. Mol Cell 15:595–605

Sempere LF, Freemantle S, Pitha-Rowe I, Moss E, Dmitrovsky E, Ambros V (2004) Expression profiling of mammalian microRNAs uncovers a subset of brain-expressed microRNAs with possible roles in murine and human neuronal differentiation. Genome Biol 5:R13

Setoguchi H, Namihira M, Kohyama J, Asano H, Sanosaka T, Nakashima K (2006) Methyl-CpG binding proteins are involved in restricting differentiation plasticity in neurons. J Neurosci Res 84:969–979

Shahbazian MD, Zoghbi HY (2002) Rett syndrome and MeCP2: linking epigenetics and neuronal function. Am J Human Genet 71:1259–1272

Shen S, Li J, Casaccia-Bonnefil P (2005) Histone modifications affect timing of oligodendrocyte progenitor differentiation in the developing rat brain. J Cell Biol 169:577–589

Stanfield BB, Trice JE (1988) Evidence that granule cells generated in the dentate gyrus of adult rats extend axonal projections. Exp Brain Res 72:399–406

Takizawa T, Nakashima K, Namihira M, Ochiai W, Uemura A, Yanagisawa M, Fujita N, Nakao M, Taga T (2001) DNA methylation is a critical cell-intrinsic determinant of astrocyte differentiation in the fetal brain. Dev Cell 1:749–758

Tang X, Falls DL, Li X, Lane T, Luskin MB (2007) Antigen-retrieval procedure for bromodeoxyuridine immunolabeling with concurrent labeling of nuclear DNA and antigens damaged by HCl pretreatment. J Neurosci 27:5837–5844

Torres-Padilla ME, Parfitt DE, Kouzarides T, Zernicka-Goetz M (2007) Histone arginine methylation regulates pluripotency in the early mouse embryo. Nature 445:214–218

Ueba T, Kaspar B, Zhao X, Gage FH (1999) Repression of human fibroblast growth factor 2 by a novel transcription factor. J Biol Chem 274:10382–10387

Visvanathan J, Lee S, Lee B, Lee JW, Lee SK (2007) The microRNA miR-124 antagonizes the anti-neural REST/SCP1 pathway during embryonic CNS development. Genes Dev 21:744–749

Wightman B, Ha I, Ruvkun G (1993) Posttranscriptional regulation of the heterochronic gene lin-14 by lin-4 mediates temporal pattern formation in C. elegans. Cell 75:855–862

Wu H, Xu J, Pang ZP, Ge W, Kim KJ, Blanchi B, Chen C, Sudhof TC, Sun YE (2007) Integrative genomic and functional analyses reveal neuronal subtype differentiation bias in human embryonic stem cell lines. Proc Natl Acad Sci U-SA 104:13821–13826

Xiao C, Calado DP, Galler G, Thai TH, Patterson HC, Wang J, Rajewsky N, Bender TP, Rajewsky K (2007) MiR-150 controls B cell differentiation by targeting the transcription factor c-Myb. Cell 131:146–159

Zhao X, Ueba T, Christie BR, Barkho B, McConnell MJ, Nakashima K, Lein ES, Eadie BD, Willhoite AR, Muotri AR, Summers RG, Chun J, Lee KF, Gage FH (2003) Mice lacking methyl-CpG binding protein 1 have deficits in adult neurogenesis and hippocampal function. Proc Natl Acad Sci USA 100:6777–6782

Zhao X, Schaffer D, Gage F (2004) Neurogenesis in adult brain: understanding the mechanism and regulation. In: Gage FH, Bjöund A, Prochiantz A, Christen Y (eds) Stem cells in the nervous system: functional and clinical implications. Heidelberg, Springer-Verlag Berlin pp 139–146

microRNAs in CNS Development and Neurodegeneration: Insights from Drosophila Genetics

Stephen M. Cohen

Abstract Age-related neurodegeneration can profoundly impair quality of life in the elderly. Although the causes of many types of familial neurodegenerative diseases are known, genetic forms of disease affect relatively few people. For most people, age is the most prominent risk factor for developing neurodegenerative disease, and very little is known about its molecular causes. Recent advances have linked small regulatory RNA molecules, called microRNAs, to neurodegenerative disorders in animal models, including the fruit fly, Drosophila. Although a fly's brain is vastly simpler than a human brain, much is similar at the level of individual cells and fly models are proving to be useful to understanding the mechanisms that underlie neurodegenerative disease. Our aim is to understand the basic cellular processes that can cause neurodegeneration in the fly model as a means of identifying new links to disease. We are engaged in a large-scale effort to systematically analyze the roles of microRNA genes in the fly brain. The comparative ease and speed with which Drosophila can be studied in the laboratory allow the possibility of a survey of its entire genome for the microRNAs that are required for proper cognitive function and the health and survival of the cells of the brain. Identification of the targets of such microRNAs may help to identify new causes of neurodegenerative disease.

1 Introduction

Neurodegenerative diseases result from progressive impairment of the function of the nerve cells in the brain. This impairment can be associated with degeneration of nerve cell structure, with the appearance of abnormal protein deposits, and with

S.M. Cohen
Temasek Life Sciences Laboratory and Department of Biological Sciences, National University of Singapore, 1 Research Link, Singapore, 117604 Singapore

B. de Strooper and Y. Christen (eds.), *Macro Roles for MicroRNAs in the Life and Death of Neurons*, Research and Perspectives in Neurosciences,
DOI 10.1007/978-3-642-04298-0_8,

progressive loss of nerve cells. These cellular defects ultimately lead to cognitive and behavioral impairment. There are many different ways in which nerve cells can lose their ability to function normally and so lead to impaired functioning of the brain. Understanding the causes of impaired nerve cell function that can lead to disease poses a major challenge.

For some neurodegenerative diseases, genetic causes are known. Purely familial disorders include those caused by dominant mutations that alter a normal gene to encode expanded polyglutamine stretches, e.g., in Huntington's disease and Spinocerebellar Ataxia type 3. In addition, rare genetic predispositions have been identified for many "sporadic" neurodegenerative disorders; carriers of these alleles can pass on the disease, and their offspring are at risk of developing an early-onset form of the disease.

In the cases where specific molecular lesions have been associated with disease, it has been possible to produce animal models. These models have proven a valuable resource to better understand the molecular origins of the disease state, to study its early onset and to explore potential therapeutic approaches (reviewed in Cole 2006; Lu and Vogel 2009; Terzioglu and Galter 2008). For example, recent genetic screens performed using fly disease models have raised the possibility that the problems associated with polyglutamine repeat expansion disease genes might lie at the level of the RNA (Li et al. 2008). This finding raises intriguing questions about whether the altered protein sequence, polyglutamine, is directly involved.

Another recent study has suggested new avenues to explore for the causes of Parkinson's disease. Gain-of-function mutations in LRRK2 have been associated with adult-onset Parkinson's disease (Abeliovich and Flint Beal 2006; West et al. 2005, 2007). Recent studies have made use of a fly model, in which a mutant form of LRRK2 found associated with human disease was expressed in the fly brain. This expression was found to cause altered dopamine levels and age-dependent loss of dopaminergic neurons but not of other neuronal cell types, suggesting that the fly model is relevant to human disease (Imai et al. 2008). These authors have shown that LRRK2 may act to regulate the activity of the TOR signaling pathway to control cellular protein biosynthetic capacity. This study raises intriguing questions about how elevated protein biosynthetic capacity compromises the survival of dopaminergic neurons, and it illustrates the potential utility of the genetically tractable fly models to provide new insight into disease mechanism.

2 microRNAs, the brain and disease

A number of recent studies have implicated microRNAs in neurodegeneration (reviewed in Bushati and Cohen 2008; Hebert and De Strooper 2009). microRNAs are small endogenous RNAs that act as post-transcriptional regulators of gene expression. microRNAs serve as guide molecules to target the microRNA-containing ribonucleoprotein complexes (miRNP) to their target messenger RNAs (mRNA). Binding of the miRNP to the target can affect expression of the target mRNA (reviewed in Bartel 2009). Typically, target levels are reduced, either by blocking translation of the target mRNA to produce protein or through destabilization as a consequence of mRNA deadenylation (reviewed in Eulalio et al. 2008; Filipowicz et al. 2008). However, instances of microRNA-mediated upregulation of targets have been documented (e.g., Orom et al. 2008; Vasudevan et al. 2007).

microRNAs are predicted to target hundreds of protein-coding genes, and so their regulatory potential is large (e.g., Krek et al. 2005; Lewis et al. 2005; Stark et al. 2005). Many microRNAs are expressed in the central nervous system, often in a temporally and/or spatially regulated manner during development (Landgraf et al. 2007; Berezikov et al. 2006; Bak et al. 2008; Kapsimali et al. 2007). To date, several hundred microRNAs have been identified in human and chimpanzee brain (Landgraf et al. 2007; Berezikov et al. 2006), and it has been estimated that this number may exceed 1,000 microRNAs expressed in human brain (Berezikov et al. 2006). Many brain microRNAs are not conserved beyond primates, suggesting a comparatively recent evolutionary origin.

Given the very large numbers of microRNAs, and the comparatively short time that we have known about them, it is not surprising that few of the brain-specific microRNAs have been assigned well-defined biological functions. Nonetheless, evidence is beginning to accumulate for roles in normal development, differentiation and function, as well as in pathology. For example, microRNA target sites are beginning to emerge as single nucleotide polymorphisms (SNP) associated with disease (Sethupathy and Collins 2008), reflecting the way that microRNAs find their targets. microRNAs provide the sequence information that guides the miRNP to target mRNAs. Base-pairing interactions involving six to eight residues at the 5' end of the microRNA, called the "seed," are thought to provide most of the target recognition specificity, with additional base-pairings contributing to the magnitude of regulation (Lewis et al. 2005; Stark et al. 2005). Target sites are frequently found in the non-coding portion of mRNAs, typically in 3' UTRs, so it is easy to imagine how polymorphisms in non-coding sequences could generate a novel target sites. Changing one residue in a 3'UTR might create a new seed match, leading to reduced expression of that allelic form of the mRNA. Similarly, loss of a microRNA site could be caused by a 3'UTR SNP, which would then lead to elevated expression of that allele. In many cases the effects would be quantitatively modest and so could be associated with enhanced or reduced disease susceptibility, without being overt disease mutations. To illustrate this possibility, a recent study identified a SNP that creates a site for human miR-659 in the 3' UTR of the progranulin gene as a risk factor in a specific form of dementia (Rademakers et al. 2008). Another recent study reported a SNP that destroys a site for miR-433 in the 3' UTR of FGF20 and is linked to an increased risk of Alzheimer's disease by indirectly causing overexpression of a-Synuclein (Wang et al. 2008). Undoubtedly, many more examples of similar relationships will emerge, given the large numbers of CNS-expressed microRNAs and the potential for subtle changes in gene expression to have profound consequences over the long term in the long-lived cells of the brain.

3 Approaches to the study of microRNA functions

The roles of microRNAs in the brain have been studied using a variety of genetic and molecular genetic approaches. One means to assess the global role of microRNAs in the brain is through use of mutations, such as *dicer*, that compromise microRNA biogenesis. Loss of all microRNAs results, causing misregulation of hundreds, if not thousands, of protein-coding genes. Although this is something of a sledgehammer approach to dissecting the functions of microRNAs in neurons, it has yielded interesting observations. Mouse *dicer* mutants die before neurulation (Bernstein et al. 2003), so a conditional knockout approach is needed. Cell type-specific removal of *Dicer* from a variety of mouse neuronal cell types has revealed defects in neuronal survival during development and in mature neurons. Depletion of all microRNAs in this way can lead to progressive loss of these cells and to behavioral defects reminiscent of the phenotypes seen in the pathologies of neurodegenerative disorders (Kim et al. 2007; Davis et al. 2008; Schaefer et al. 2007; Choi et al. 2008; Damiani et al. 2008). A better understanding of which microRNAs are responsible for these problems and identification of their biologically important targets are the next goals. One approach involves restoring specific microRNAs in dicer-depleted embryos. Giraldez et al. (2005) successfully pioneered this approach in zebrafish, showing that the miR-430 family played an important early role in CNS morphogenesis.

A more specific approach involves exploring the function of individual microRNAs. There are two general strategies: depletion of individual microRNAs using anti-sense oligonucleotides and removal of individual microRNAs by generating loss of function mutations. microRNA depletion can be effective in experimental models that are accessible to transfection or electroporation, provided that adequate controls are done for non-specific neurotoxicity and off-target effects. Injection of antisense oligonucleotides into zebrafish embryos has been used to document a role for the miR-200 family of microRNAs in differentiation of olfactory neuronal progenitor cells (Choi et al. 2008). In cultured hippocampal neurons, miR-134 was shown by antisense-mediated depletion to be involved in dendritic spine morphology by regulating Limk1 expression (Schratt et al. 2006).

Depletion of miR-124 has recently been reported to reduce or delay differentiation of mouse stem cell progenitors in culture and in the developing subventricular zone of the mouse brain (Cheng et al. 2009), although earlier reports did not observe such effects in electroporated chick neural tubes (Cao et al. 2007). Conversely, miR-124 overexpression can promote neuronal differentiation in vivo and in a variety of neural progenitor cells types in vitro (Cao et al. 2007; Cheng et al. 2009; Makeyev et al. 2007; Visvanathan et al. 2007). Different target mRNAs have been implicated in this process in each of these studies. Analysis of miR-124 mutants may help to assess the biological roles of these candidate targets in vivo. However, mouse and humans have three separate miR-124 genes, making a genetic analysis complex.

In genetically tractable model systems, production of mutations that remove microRNA function can be informative. In *C. elegans*, microRNAs are required for specification and maintenance of gustatory sensory neurons (Johnston et al. 2005).

In *Drosophila*, miR-7 maintains the differentiated state of photoreceptors (Li and Carthew 2005) and miR-9 is involved in specification of sense organ precursors (Li et al. 2006). The variety of phenotypes uncovered in these studies hints at the diversity of microRNA functions in the central nervous system.

4 microRNAs and neurodegenerative disease

A variety of different approaches are beginning to implicate microRNAs in biological processes relevant to neurodegenerative disease. Expansion of the polyQ repeats in a variety of proteins is linked to neurodegeneration (reviewed in Orr and Zoghbi 2007). Depletion of *dicer* in HeLa cells significantly enhanced pathogenic Ataxin-3-induced toxicity (Bilen et al. 2006), which could be partially rescued by complementing the cells with the purified small RNA fraction containing total HeLa cell microRNAs and using an in vivo polyQ-disease fly model by overexpression of the bantam microRNA (Bilen et al. 2006), which has been shown to be anti-apoptotic (Brennecke et al. 2003). The polyQ disorder SCA1 is characterized by the death of cerebellar Purkinje cells (reviewed in Orr and Zoghbi 2007). Depletion of *dicer* from mouse Purkinje neurons did not impair cell survival in young mice, but, by 13 weeks of age, Purkinje neurons began to degenerate and were lost by apoptosis. These mice developed mild ataxia, which became more severe with age (Schaefer et al. 2007). As in the case of SCA3, the microRNAs seem to have a protective function. Whether specific microRNAs are affected in patients with these diseases remains to be determined.

microRNAs have also been associated with loss of dopaminergic neurons in Parkinson's disease. miR-133b was found to be expressed at reduced levels in Parkinson's disease brains and in animal models of the disease (Kim et al. 2007). Overexpression of miR-133 inhibited differentiation of embryonic stem cells into dopaminergic neurons, whereas its depletion using antisense oligonucleotides increased expression of dopaminergic neuronal markers. The authors identified the transcription factor pitx3 as a target of miR-133b and provided evidence that pitx3 in turn regulates miR-133b expression, providing a feedback loop required for normal control of gene expression in this important cell type. Whether misregulation of miR-133b is linked to clinical disease remains to be determined.

The polyQ repeat of human Atrophin-1 is expanded in patients with dentatorubral-pallidoluysian atrophy (DRPLA), resulting in neuronal apoptosis. Analysis of the fly microRNA miR-8 has implicated Drosophila Atrophin orthologue in neurodegeneration. miR-8 sets the levels of Atrophin expression by binding to its 3'UTR. Flies lacking *miR-8* perform poorly in an assay for motor coordination and their performance declines more rapidly with age than in normal flies. They also show an abnormal level of cell death in the brain. These defects can be ameliorated by limiting the degree to which Atrophin can be overexpressed in the microRNA mutant, indicating that elevated Atrophin levels are responsible for the neurodegenerative defects (Karres et al. 2007). Is there a comparable relationship between a microRNA and human Atrophin-1? miR-8 is related to mouse and human miR-200b and miR-429, which can target the Atrophin orthologue RERE (Karres et al. 2007). RERE has been shown to bind Atrophin-1 (Yanagisawa et al. 2000) and to induce apoptosis upon overexpression in neuronal cells in culture (Waerner et al. 2001). Whether preventing microRNA-mediated repression of RERE would cause similar defects in the mouse brain remains to be determined. It would also be of interest to know whether levels of miR-200b and miR-429 are affected in DRPLA patients or if SNPs affecting their ability to regulate RERE might correlate with disease.

5 A genome-wide screen for Drosophila microRNAs linked to normal CNS function and neurodegeneration

Although a fly's brain is simple, fly neurons as individual cells are quite similar to those in humans. Our experience with studying miR-8 made us aware of the opportunity to use microRNA mutants to explore the cell biology of neurodegeneration. The ease and speed with which Drosophila mutants can be generated, using homologous recombination-mediated gene targeting, allows us to survey the complete set of fly microRNAs for those that are required for the health and survival of the cells of the brain, as well as for those required for proper cognitive function and behavior. A systematic microRNA gene knock-out effort is ongoing in the author's laboratory to generate these mutants (Chen, Weng, Verma, Hilgers, Bushati, Varghese and Cohen, unpublished observations).

Deep sequencing efforts have provided evidence that as many as two thirds of all fly microRNAs may be expressed in the adult head: 44 of the 152 identified fly microRNAs have more than 10% of their total expression in the adult head (Ruby et al. 2007). To date, mutants have been reported in only a handful of these, including miR-8 (discussed above; Karres et al. 2007), the let-7, miR100 125 cluster (Sokol et al. 2008) and miR-7 (Li and Carthew 2005). miR-7 has been shown to ensure robustness of retinal neuronal cell fate specification. The let-7 cluster is expressed in larval and adult nervous system and muscles, and mutants display behavioral defects. Our preliminary analysis of the newly generated microRNA mutants suggests that behavioral defects will be an emerging theme among CNS-expressed microRNAs. As well, we see evidence for roles for specific microRNAs in CNS neuronal progenitor cell proliferation, neuroendocrine control of metabolism, etc. (Chen, Weng, Verma, Hilgers, Bushati, Varghese and Cohen, unpublished observations).

6 Prospects

To date, it remains uncertain whether microRNAs are actively involved in the etiology or progression of neurodegenerative disorders. microRNAs seem to be required for the survival of specific types of mature neurons, but whether loss of individual microRNAs can account for disease remains to be determined. Each microRNA can regulate hundreds of target genes, and so each microRNA expressed in the brain has the potential to cause a considerable shift in neuronal gene expression. Much of this will likely be innocuous, but some targets may prove to be causally linked to pathology. Similarly the acquisition of loss of microRNA target sites due to polymorphisms may prove to be important in disease. A systematic approach to identifying microRNA targets linked with neuropathology in a simple model organism, the fly, may shorten the path to discovery of disease-associated genes and may lead to new avenues for diagnosis and therapy.

Acknowledgments I thank Natascha Bushati, Yawen Chen, Valerie Hilgers, Xin Hong, Jishy Varghese, Pushpa Verma and Ruifen Weng for their contributions to the unpublished work discussed here. Work in the author's lab is supported by EU-FP6 grant "Sirocco" LSHG-CT-2006-037900, the Temasek Life Sciences Laboratory and the National Research Foundation of Singapore.

References

Abeliovich A, Flint Beal M (2006) Parkinsonism genes: culprits and clues. J Neurochem 99:1062–1072

Bak M, Silahtaroglu A, Moller, M, Christensen M, Rath MF, Skryabin B, Tommerup N, Kauppinen, S (2008) MicroRNA expression in the adult mouse central nervous system. RNA 14:432–444

Bartel DP (2009) MicroRNAs: target recognition and regulatory functions. Cell 136:215–233

Berezikov E, Thuemmler F, van Laake LW, Kondova I, Bontrop R, Cuppen E, Plasterk RH (2006) Diversity of microRNAs in human and chimpanzee brain Nature Genet 38:1375–1377

Bernstein E, Kim SY, Carmell MA, Murchison EP, Alcorn H, Li MZ, Mills AA, Elledge SJ, Anderson KV, Hannon GJ (2003) Dicer is essential for mouse development. Nature Genet 35:215–217

Bilen J, Liu N, Burnett BG, Pittman RN, Bonini NM (2006) MicroRNA pathways modulate polyglutamine-induced neurodegeneration. Mol Cell 24:157–163

Brennecke J, Hipfner DR, Stark A, Russell RB, Cohen SM (2003) bantam encodes a developmentally regulated microRNA that controls cell proliferation and regulates the pro-apoptotic gene hid in Drosophila. Cell 113:25–36

Bushati N, Cohen SM (2008) microRNAs in neurodegeneration. Curr Opin Neurobiol 8:292–296

Cao X, Pfaff SL, Gage FH (2007) A functional study of miR-124 in the developing neural tube. Genes Dev 21:531–536

Cheng LC, Pastrana E, Tavazoie M, Doetsch F (2009) miR-124 regulates adult neurogenesis in the subventricular zone stem cell niche. Nature Neurosci 12:399–408

Choi PS, Zakhary L, Choi WY, Caron S, Alvarez-Saavedra E, Miska EA, McManus M, Harfe B, Giraldez AJ, Horvitz RH, Schier AF, Dulac C (2008) Members of the miRNA-200 family regulate olfactory neurogenesis. Neuron 57:41–55

Cole G (2006) A transgenic triple scores a home run. Nature Med 12:762–763

Damiani D, Alexander JJ, O'Rourke JR, McManus M, Jadhav AP, Cepko CL, Hauswirth WW, Harfe BD, Strettoi E (2008) Dicer inactivation leads to progressive functional and structural degeneration of the mouse retina. J Neurosci 28:4878–4887

Davis TH, Cuellar TL, Koch SM, Barker AJ, Harfe BD, McManus MT, Ullian EM (2008) Conditional loss of Dicer disrupts cellular and tissue morphogenesis in the cortex and hippocampus. J Neurosci 28:4322–4330

Eulalio A, Huntzinger E, Izaurralde E (2008) Getting to the root of miRNA-mediated gene silencing. Cell 132:9–14

Filipowicz W, Bhattacharyya SN, Sonenberg N (2008) Mechanisms of post-transcriptional regulation by microRNAs: are the answers in sight? Nature Rev Genet 9:102–114

Giraldez AJ, Cinalli RM, Glasner ME, Enright AJ, Thomson JM, Baskerville S, Hammond SM, Bartel DP, Schier AF (2005) MicroRNAs regulate brain morphogenesis in zebrafish. Science 308:833–838

Hebert SS, De Strooper B (2009) Alterations of the microRNA network cause neurodegenerative disease. Trends Neurosci 32:199–206

Imai Y, Gehrke S, Wang HQ, Takahashi R, Hasegawa K, Oota E, Lu B (2008) Phosphorylation of 4E-BP by LRRK2 affects the maintenance of dopaminergic neurons in Drosophila EMBO J. 27:2432–2443

Johnston RJ, Jr, Chang S, Etchberger JF, Ortiz CO, Hobert O (2005) MicroRNAs acting in a double-negative feedback loop to control a neuronal cell fate decision. Proc Natl Acad Sci USA 102:12449–12454

Kapsimali M, Kloosterman WP, de Bruijn E, Rosa F, Plasterk RH, Wilson SW (2007) MicroRNAs show a wide diversity of expression profiles in the developing and mature central nervous system. Genome Biol 8:R173

Karres JS, Hilgers V, Carrera I, Treisman J, Cohen SM (2007) The conserved microRNA miR-8 tunes atrophin levels to prevent neurodegeneration in Drosophila. Cell 131:136–145

Kim J, Inoue K, Ishii J, Vanti WB, Voronov SV, Murchison E, Hannon G, Abeliovich A (2007) A MicroRNA feedback circuit in midbrain dopamine neurons. Science 317:1220–1224

Krek A, Grun D, Poy MN, Wolf R, Rosenberg L, Epstein EJ, MacMenamin P, da Piedade I, Gunsalus KC, Stoffel M, Rajewsky N (2005) Combinatorial microRNA target predictions. Nature Genet 37:495–500

Landgraf P, Rusu M, Sheridan R, Sewer A, Iovino N, Aravin A, Pfeffer S, Rice A, Kamphorst AO, Landthaler M, Lin C, Socci ND, Hermida L, Fulci V, Chiaretti S, Foà R, Schliwka J, Fuchs U, Novosel A, Müller RU, Schermer B, Bissels U, Inman J, Phan Q, Chien M, Weir DB, Choksi R, De Vita G, Frezzetti D, Trompeter HI, Hornung V, Teng G, Hartmann G, Palkovits M, Di Lauro R, Wernet P, Macino G, Rogler CE, Nagle JW, Ju J, Papavasiliou FN, Benzing T, Lichter P, Tam W, Brownstein MJ, Bosio A, Borkhardt A, Russo JJ, Sander C, Zavolan M, Tuschl T (2007) A mammalian microRNA expression atlas based on small RNA library sequencing. Cell 129:1401–1414

Lewis BP, Burge CB, Bartel DP (2005) Conserved seed pairing, often flanked by adenosines, indicates that thousands of human genes are microRNA targets. Cell 120:15–20

Li LB, Yu Z, Teng X, Bonini NM (2008) RNA toxicity is a component of ataxin3 degeneration in Drosophila. Nature 453:1107–1111

Li X, Carthew RW (2005) A microRNA mediates EGF receptor signaling and promotes photoreceptor differentiation in the Drosophila eye. Cell 123:1267–1277

Li Y, Wang F, Lee JA, Gao FB (2006) MicroRNA-9a ensures the precise specification of sensory organ precursors in Drosophila. Genes Dev 20:2793–2805

Lu B, Vogel H (2009) Drosophila models of neurodegenerative diseases. Annu Rev Pathol 4:315–342

Makeyev EV, Zhang J, Carrasco MA, Maniatis T (2007) The MicroRNA miR-124 promotes neuronal differentiation by triggering brain-specific alternative pre-mRNA splicing. Mol Cell 27:435–448

Orom UA, Nielsen FC, Lund AH (2008) MicroRNA-10a binds the 5'UTR of ribosomal protein mRNAs and enhances their translation. Mol Cell 30:460–471

Orr HT, Zoghbi HY (2007) Trinucleotide repeat disorders. Annu Rev Neurosci 30:575–621

Rademakers R, Eriksen JL, Baker M, Robinson T, Ahmed Z, Lincoln SJ, Finch N, Rutherford NJ, Crook RJ, Josephs KA, Boeve BF, Knopman DS, Petersen RC, Parisi JE, Caselli RJ, Wszolek ZK, Uitti RJ, Feldman H, Hutton ML, Mackenzie IR, Graff-Radford NR, Dickson DW (2008) Common variation in the miR-659 binding-site of GRN is a major risk factor for TDP43-positive frontotemporal dementia. Human Mol Genet 17:3631–3642

Ruby JG, Stark A, Johnston WK, Kellis M, Bartel DP, Lai EC (2007) Evolution, biogenesis, expression and target predictions of a substantially expanded set of Drosophila microRNAs. Genome Res 17:1850–1864

Schaefer A, O'Carroll D, Tan CL, Hillman D, Sugimori M, Llinas R, Greengard P (2007) Cerebellar neurodegeneration in the absence of microRNAs. J Exp Med 204:1553–1558

Schratt GM, Tuebing F, Nigh EA, Kane CG, Sabatini ME, Kiebler M, Greenberg ME (2006) A brain-specific microRNA regulates dendritic spine development. Nature 439:283–289

Sethupathy P, Collins FS (2008) MicroRNA target site polymorphisms and human disease. Trends Genet 24:489–497

Sokol NS, Xu P, Jan YN, Ambros V (2008) Drosophila let-7 microRNA is required for remodeling of the neuromusculature during metamorphosis. Genes Dev 22:1591–1596

Stark A, Brennecke J, Bushati N, Russell RB, Cohen SM (2005) Animal MicroRNAs confer robustness to gene expression and have a significant impact on 3'UTR evolution. Cell 123:1133–1146

Terzioglu M, Galter D (2008) Parkinson's disease: genetic versus toxin-induced rodent models. FEBS J 275:1384–1391

Vasudevan S, Tong Y, Steitz JA (2007) Switching from repression to activation: microRNAs can up-regulate translation. Science 318:1931–1934

Visvanathan J, Lee S, Lee B, Lee JW, Lee SK (2007) The microRNA miR-124 antagonizes the anti-neural REST/SCP1 pathway during embryonic CNS development Genes Dev 21:744–749

Waerner T, Gardellin P, Pfizenmaier K, Weith A, Kraut N (2001) Human RERE is localized to nuclear promyelocytic leukemia oncogenic domains and enhances apoptosis. Cell Growth Differ 12:201–210

Wang G, van der Walt JM, Mayhew G, Li YJ, Zuchner S, Scott WK, Martin ER and Vance JM (2008) Variation in the miRNA-433 binding site of FGF20 confers risk for Parkinson disease by overexpression of alpha-synuclein. Am J Human Genet 82:283–289

West AB, Moore DJ, Biskup S, Bugayenko A, Smith WW, Ross CA, Dawson VL, Dawson TM (2005) Parkinson's disease-associated mutations in leucine-rich repeat kinase 2 augment kinase activity. Proc Natl Acad Sci USA 102:16842–16847

West AB, Moore DJ, Choi C, Andrabi SA, Li X, Dikeman D, Biskup S, Zhang Z, Lim KL, Dawson VL, Dawson TM (2007) Parkinson's disease-associated mutations in LRRK2 link enhanced GTP-binding and kinase activities to neuronal toxicity. Human Mol Genet 16:223–232

Yanagisawa H, Bundo M, Miyashita T, Okamura-Oho Y, Tadokoro K, Tokunaga K, Yamada M (2000) Protein binding of a DRPLA family through arginine-glutamic acid dipeptide repeats is enhanced by extended polyglutamine. Human Mol Genet 9:1433–1442

Drosophila as a Model for Neurodegenerative Disease: Roles of RNA Pathways in Pathogenesis

Nancy M. Bonini

Abstract Human neurodegenerative diseases, including the polyglutamine diseases like Huntington's disease and Spinocerebellar ataxia type 3, are late-onset progressive neurodegenerative disorders for which few cures or treatments are available. To develop new approaches to the understanding of and foundation for therapeutics for these diseases, we are using *Drosophila* as a model. In these studies, we have expressed the human disease gene for the polyglutamine disorder Spinocerebellar ataxia type 3 in flies. Expression of a normal, non-mutant protein with a short polyglutamine stretch has no effect, but expression of a disease form of the protein with an expanded polyglutamine confers late-onset progressive neurodegeneration.

Our research has subsequently focused on modifier screens for genes that modulate the neurodegeneration. These genetic screens have confirmed our earlier findings that modulation of protein solubility and protein quality control pathways has a dramatic effect on polyglutamine protein toxicity. In our genetic screens, however, we also found evidence of a role for microRNAs in disease and a pathogenic role of the CAG repeat RNA in polyglutamine-induced degeneration. For the microRNA pathway, we found that the microRNA *bantam* is a dosage-sensitive modifier of polyglutamine toxicity and that compromise of the entire microRNA pathway is more severely deleterious. In other screens, we isolated the gene *muscleblind* as an upregulation enhancer of polyglutamine protein toxicity. This finding suggested to us that there might be a role for the CAG-expanded repeat RNA in polyglutamine degeneration because *muscleblind* is a known modulator of CUG-based RNA diseases. We further tested the role of the RNA in polyglutamine toxicity by altering the CAG repeat sequence to an interrupted CAA/CAG repeat within the polyglutamine-encoding region; this alteration mitigated

N.M. Bonini
Department of Biology, University of Pennsylvania, Howard Hughes Medical Institute, 306 Leidy Laboratories, Philadelphia, PA 19104-6018, USA
e-mail: nbonini@sas.upenn.edu

B. de Strooper and Y. Christen (eds.), *Macro Roles for MicroRNAs in the Life and Death of Neurons*, Research and Perspectives in Neurosciences,
DOI 10.1007/978-3-642-04298-0_9, © Springer-Verlag Berlin Heidelberg 2010

toxicity. We further found that expression of an untranslated CAG repeat of pathogenic length confered neural dysfunction and degeneration. These studies indicate that a number of different RNA pathways modulate polyglutamine disease and neurodegeneration, including microRNA pathways and, for CAG repeat diseases, the RNA itself. This latter finding highlights common components between RNA-based and polyglutamine protein-based repeat expansion diseases.

1 Introduction

The polyglutamine diseases are a set of human diseases caused by the expansion of a CAG-repeat encoding the amino acid glutamine within the open reading frames of the respective genes. These diseases include Huntington's disease and a number of ataxias like spinocerebellar ataxia types 1 (SCA1), SCA2 and SCA3. In all, there are nine currently known polyglutamine diseases (Orr and Zoghbi 2007; Williams and Paulson 2008).

The polyglutamine diseases are a subclass of a much larger set of human diseases known as the trinucleotide repeat diseases, all due to small repeat expansions, typically of triplet repeats within genes (Pearson et al. 2005). The repeats can occur in different regions of the genes in the different diseases, such as in the 5′UTR, 3′UTR or introns. For the polyglutamine diseases, the repeat occurs within the coding region. Thus the polyglutamine disease proteins normally contain a polyglutamine stretch; in disease, the polyglutamine domain becomes expanded to be abnormally long. The polyglutamine expansion is generally thought to confer dominant toxicity on the disease protein, leading to neuronal dysfunction and loss. The pathogenic polyglutamine protein undergoes an abnormal aggregation process to form accumulations, typically nuclear inclusions, in the cells. Although it is unclear if the large inclusions are toxic, abnormal aggregation of the protein is correlated with toxicity and disease.

To develop new approaches to the understanding and foundation for therapeutics of these diseases, we and others are using *Drosophila* as a model (Bilen and Bonini 2005; Marsh and Thompson 2006). *Drosophila* has many pathways conserved to humans and has a complex nervous system with complex behaviors. In our studies, we have expressed the human disease gene for the polyglutamine disorder SCA3 in flies. Expression of a normal, non-mutant Ataxin-3 protein with a short polyglutamine stretch has no effect, but expression of a disease form of the Ataxin-3 protein with an expanded polyglutamine confers late-onset progressive neurodegeneration (Warrick et al. 1999).

With a *Drosophila* model in hand, a variety of powerful tools are available in the fly to learn about how the disease protein causes degeneration and to discover ways to interfere with the disease processes. *Drosophila* has a wealth of experimental genetic approaches that can be brought to bear on the disease processes and has been a powerful discovery agent for defining genes critical to maintenance of the nervous system that are conserved in humans (Bier 2005; Lessing and Bonini 2009). With a disease model in hand, one can then apply fly genetics to find modifier pathways that enhance or suppress the disease situation. In these studies, we found that molecular chaperones are powerful modulators of the disease phenotype (Bonini 2002; Warrick et al.1999). Indeed molecular chaperones are dosage-sensitive modifiers, with upregulation of chaperones leading to suppression and downregulation of molecular chaperones resulting in enhancement. This finding indicates that the disease effect is very sensitive to the levels of molecular chaperones.

We extended this finding of a role for molecular chaperones to other disease situations in the fly, notably a model for Parkinson's disease. Dominant Parkinson's disease is also associated with the abnormal accumulation of protein, in this case the small synaptic protein alpha-synuclein (Lee and Trojanowski 2006; Selkoe 2004). Expression of alpha-synuclein in flies leads to a decrease in tyrosine hydroxylase immunostaining of dopaminergic neurons and to the accumulation of alpha-synuclein in the Lewy body-like aggregates that characterize human Parkinson's disease. Upregulation of the molecular chaperone Hsp70 mitigates alpha-synuclein toxicity in flies (Auluck et al. 2002). Hsp70 is also co-localized with Lewy bodies in human diseases associated with alpha-synuclein accumulation, and a number of studies now indicate that Hsp70 can modulate alpha-synuclein aggregation and oligomers (Dedmon et al. 2005; Zhou et al. 2004). Importantly, the presence of polymorphisms in one of the Hsp70 genes is a risk factor for human Parkinson's disease; the polymorphism compromises upregulation of Hsp70 in cells in culture in response to stress (Wu et al. 2004). Taken together, these studies emphasize the critical importance of protein quality control pathways in age-dependent disease effects and underscore the value of the fly in revealing insight into features of disease relevant to humans.

For polyglutamine disease, we have performed a number of genetic screens to define genes that modulate neurodegeneration. One of our largest screens defined a variety of gene products that mitigated neuronal loss and dysfunction (Bilen and Bonini 2007). A challenge of the screens is to define the mechanisms by which the modifiers function; to do this, we focused on several features. These features include what the consequence is to the accumulation of the disease protein, the solubility of the disease protein, and the extent to which modifiers affect the toxicity of other disease proteins or other potentially related effects. Using these approaches, we found that many modifiers that can mitigate polyglutamine toxicity can also mitigate the effect of massively compromising Hsp70 molecular chaperone activity. This finding suggested to us that these modifiers ultimately biologically integrate onto protein quality control pathways to effect rescue (Fig. 1). However, among these modifiers, we also found select modifiers that appeared to act in different ways. We also focused on these modifiers to define additional pathways that modulate the toxic disease gene effects. Two of these modifiers lead us to the idea that pathways of RNA gene regulation and RNA metabolism integrate into disease.

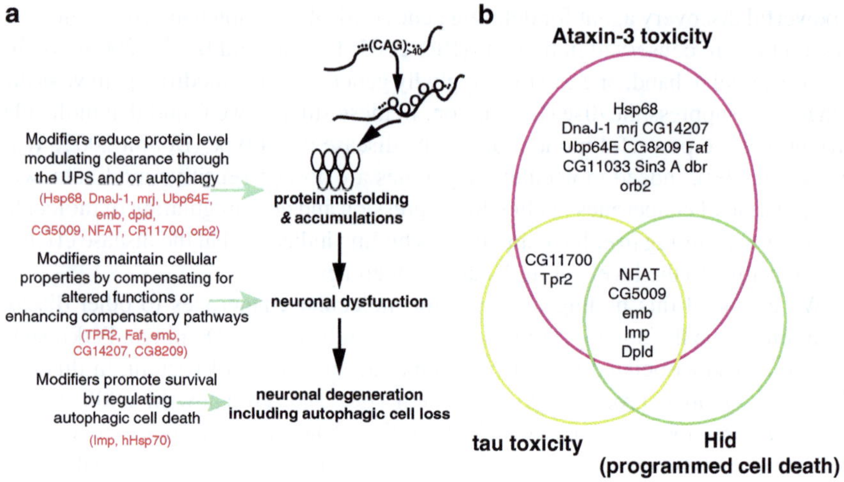

Fig. 1 Activities of modifier genes of Ataxin-3 neurodegeneration isolated in a genome-wide screen. **a.** Modifiers of pathogenic Ataxin-3 toxicity may 1) reduce disease protein accumulation into inclusions in a manner sensitive to proteasome activity, and/or by modulating autophagy, 2) promote cellular function in situations of misfolded protein and/or 3) promote neuronal survival by regulating autophagic cell loss. **b.** Select modifiers of Ataxin-3 toxicity reveal genetic links to tau toxicity and *hid*-induced programmed cell death. Some modifiers suppress both Ataxin-3 and tau toxicity, implicating chaperone and ubiquitin proteasome activity as modifiers of tau effects associated with Alzheimer's disease. Modifiers with similar effects on tau and *hid*-induced programmed cell death may be modulating tau by modulating cell death; however, these modifiers likely act in a distinct way on Ataxin-3-associated degeneration (from Bilen and Bonini 2007)

2 A role of microRNAs in polyglutamine toxicity

In our screen for modifiers of polyglutamine toxicity, we isolated an upregulation allele of the microRNA *bantam* (Bilen et al. 2006a, b; Fig. 2). An initial approach with any modifier is to define whether the modifier functions in a dose-dependent manner. Thus, we sought to determine whether *bantam* functioned as a modifier with both upregulation and loss of function. Indeed, whereas upregulation of *bantam* suppressed polyglutamine toxicity, downregulation of *bantam* activity enhanced polyglutamine toxicity, indicating that *bantam* activity is critical to polyglutamine toxicity.

Bantam is a microRNA and thus encodes a small regulatory RNA that affects the expression of select transcripts with *bantam* target sites (Brennecke et al. 2003). The known targets of *bantam* in development include cell death genes and cell growth genes. We tested whether known targets of *bantam* could modulate polyglutamine toxicity; however, these studies revealed that none of the known targets of *bantam* had an effect (Bilen et al. 2006b). Presumably, *bantam* is regulating the activity of select target genes and, through modulation of their protein expression, is affecting the disease protein toxicity. An interesting feature of *bantam* modulation of the disease protein toxicity is that there is no change in the onset or size of

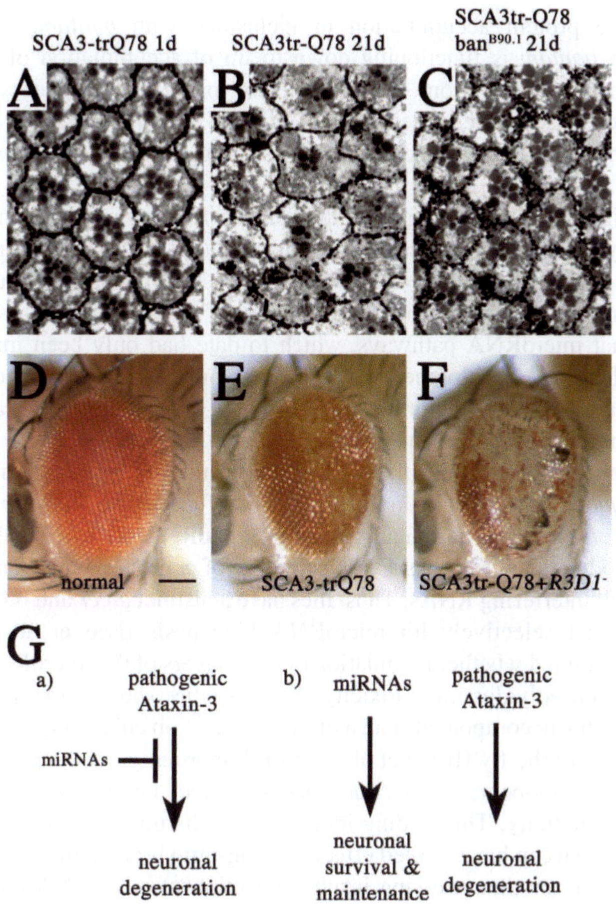

Fig. 2 Modulation of Ataxin-3 pathogenesis by the microRNA *bantam* and the microRNA pathway. **A–C.** Tangential sections of adult fly eyes. (**A**) Normally the fly eye has a highly regular pattern. (**B**) By 21 days, Ataxin-3 has induced severe degeneration, such that the eye structure is quite disrupted. (**C**) Upregulation of the microRNA *bantam* results in significantly reduced deterioration of the retinal structure by the pathogenic Ataxin-3 at 21 days. **D–F**. Compromise of microRNA processing *loquacious* mutation enhances degeneration by the Ataxin-3 protein. (**D**) A normal fly eye. (**E**) Pathogenic Ataxin-3 causes a mildly degenerate eye, reflected in disrupted pigmentation. (**F**) Reduction of microRNA processing results in dramatically enhanced degeneration, seen here as dramatically increased loss of external pigmentation. **G.** Models for how the microRNA (miRNA)pathway influences Ataxin-3-induced degeneration. **a**) microRNAs may modulate mRNA target genes that affect the toxicity of the Ataxin-3 protein, thereby mitigating neurodegeneration. One such microRNA in *Drosophila* is *bantam*. **b**) microRNAs, including *bantam* in *Drosophila*, may modulate mRNA target genes that influence neuronal survival or maintenance. Disregulation of microRNA targets may promote neuronal degeneration, including enhancing degeneration induced by pathogenic disease proteins like Ataxin-3 (from Bilen et al. 2006a)

polyglutamine protein accumulation in inclusions with *bantam*. This finding suggests that *bantam* is functioning downstream of accumulation of the protein to prevent neurodegeneration, for example, by preventing degeneration of the neurons in response to the protein insult. As known *bantam* targets could not be implicated in polyglutamine toxicity, these data suggest that novel *bantam* targets are involved.

Despite the intriguing biological implications of *bantam* activity in polyglutamine toxicity, defining biologically relevant targets of microRNAs remains challenging. We therefore addressed the more global and fundamental question that these studies of *bantam* raised: is *bantam* the tip of an iceberg? That is, is *bantam* indicating that microRNA pathways, which to date had only been implicated as regulators of normal development and cancer (Ambros 2004), may contribute to long-term maintenance of the brain and to diseases of slow onset critical to aging?

To test this idea, we went upstream of *bantam* in the microRNA pathway to modulate the activity of genes that are critical to the production of microRNAs, *dicer* and *loquacious* (an RNA binding protein that functions with Dicer to generate microRNAs). *Drosophila* has an advantage in microRNA pathway analysis in that genes that modulate microRNAs are a distinct set of genes separate from those that generate small interfering RNAs. Thus, flies have a distinct *dicer* and other pathway components that selectively hit microRNA biogenesis (Lee et al. 2004). We therefore determined whether modulation of key players of the microRNA pathway had an effect on polyglutamine toxicity. These studies showed that disruption of microRNA pathway components had a dramatic effect on enhancing polyglutamine toxicity in vivo in the fly (Bilen et al. 2006b). Importantly, modulation of at least one of these components, *loquacious*, had no effect on its own but enhanced polyglutamine toxicity. This finding indicates that disruption of microRNA pathways on their own can have limited effects in long-term brain maintenance but that, in the context of another stressing insult, disruption of microRNA pathways can have a dramatic effect.

We extended these studies to the human situation by using human cells in culture expressing normal or toxic polyglutamine protein and by disrupting *dicer* activity (Bilen et al. 2006b). In HeLa cells in culture, the normal ataxin-3 protein, encoded by the SCA3 gene, has no effect on the viability of the cells, whereas the toxic ataxin-3 protein with a long polyglutamine repeat of Q72 has some effects on causing cell death. However, with knockdown of *dicer* activity, there is a dramatic enhancement of the pathogenic ataxin-3 toxicity, such that the cells are all lost within several days, yet controls show no effect of *dicer* loss on its own or with the normal ataxin-3 protein in the same time frame. We further showed that transfection back into cells of the small RNA fraction mitigated the enhanced toxicity due to *dicer* disruption, indicating that the effect is due to small RNAs like microRNAs. These studies indicated that microRNAs have a dramatic effect on modulating the toxicity of neurodegenerative disease proteins both in vivo in the fly nervous system and in human cells in culture (Fig. 2).

These studies have been extended to confirm roles of specific microRNAs or the microRNA network in maintenance of the brain and in disease situations. Notably, cerebellar neurons, the target network of many distinct human ataxias, is sensitive to the loss of the microRNA pathway in mice (Schaefer et al. 2007). Specific microRNAs have been implicated in select neurodegenerative disease situations, including defining specific microRNA signatures in degenerative disease situations (Hebert et al. 2008; Saba et al. 2008). Thus, microRNA pathways are emerging as players that are as important to long-term brain maintenance as to development. Important future goals include the significance of changes in microRNAs on the gene networks that surround and are critical for disease proteins, as well as how microRNA regulated targets impinge on protein quality control pathways, which are by now well known to be critical to neurodegenerative disease situations and aging.

3 A pathogenic role for the CAG-repeat encoding mRNA in polyglutamine disease

The beauty of genetic screens is that, using such approaches, one can identify modulating pathways that one never anticipated. In genetic screens for an enhancer of polyglutamine toxicity, we isolated upregulation alleles of the gene *muscleblind* (Li et al. 2008). This finding was unexpected because, although *muscleblind* homologues are known in mammals as modifiers of trinucleotide repeat diseases, they are considered to be modifiers of a distinct class: CUG expansion diseases like myotonic dystrophy type I (Ranum and Cooper 2006; Wheeler and Thornton 2007). That class of triplet repeat diseases is due to toxicity at the RNA level and not protein. Thus, the isolation of *muscleblind* as a modulator of a protein toxicity was not expected, meaning that study of such a modifier may lead to new insights into polyglutamine-induced toxicity.

Study of the modulation of *muscleblind* as a modifier of polyglutamine toxicity suggested that modulation was due at least in part to the effect of *muscleblind* on upregulating the level of the polyglutamine protein. However, in addition to upregulating the protein, *muscleblind* appeared to upregulate the level of the RNA. Given that the effects of *muscleblind* on polyglutamine toxicity may therefore be due at least in part to an effect at the level of the RNA, this finding raised the more fundamental question of whether the CAG-repeat containing RNA had toxicity that contributed to polyglutamine disease.

To address this question, we performed two types of experiments. First, the toxicity of such long repeat RNAs is thought to lie in the secondary structure of the RNAs to form hairpins and, by that mechanism, potentially titrate RNA-binding proteins or perhaps trigger signaling pathways (Kuyumcu-Martinez et al. 2007; Miller et al. 2000; Sobczak et al. 2003). Thus, interruption of the hairpin structure might be anticipated to mitigate the toxicity of the RNA. We therefore interrupted the repeat sequence of the RNA from a pure CAG repeat to an interrupted CAA/CAG repeat, which is predicted by RNA folding programs (Zuker 2003) to form a distinct structure. Interruption of the repeat to disrupt the hairpin dramatically mitigated toxicity (Fig. 3). Despite the fact that the protein produced from the interrupted hairpin accumulated at the same rate and same level, including identical nuclear inclusions, the toxicity curve was shifted such that expression of the transgene, although still conferring toxicity, was mitigated. This finding indicated that expression of a protein of an identical amino acid sequence, but from an RNA no longer bearing a pure CAG repeat, was less toxic.

If a CAG repeat containing RNA can confer toxicity, then it should also be able to do so in the absence of coding for a protein. Therefore, we also made a construct that expressed a long CAG repeat RNA sequencing in the 3'UTR of a control protein DsRed. Expression of this RNA was toxic in the nervous system, conferring neuronal dysfunction and loss of integrity of the brain with time (Fig. 4). Overall the effects of the toxic RNA require a longer repeat length (and are extreme with a repeat length of 250 units compared to 100 units), and are less severe than with the protein. Moreover, the effects on non-neuronal cells of the RNA are less than with the protein. For example, the pathogenic polyglutamine protein confers an extreme toxicity to the external eye of the fly, which is composed of pigment cells, whereas the RNA does not have such an eye effect but rather, when expressed in the nervous system, causes loss of climbing ability and shortened life span. Taken together, these studies indicate that there may be a contribution of the CAG-repeat containing RNA to polyglutamine-induced neurodegeneration.

There is increasing evidence of potential contributions of CAG repeat RNA to non-polyglutamine trinucleotide repeat diseases as well from other data.

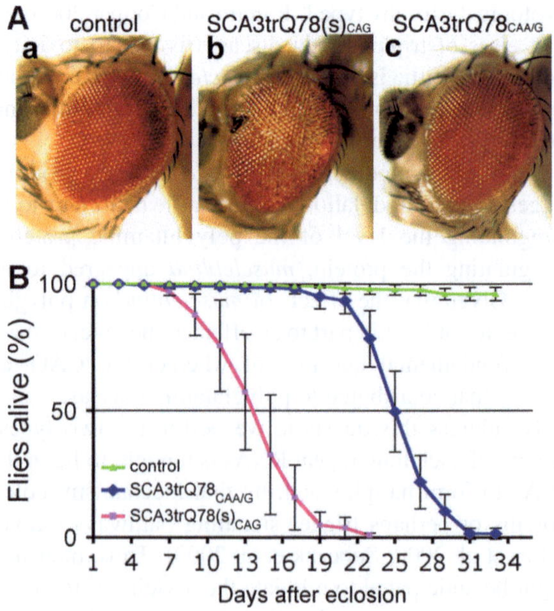

Fig. 3 Interruptions of the CAG repeat mitigate SCA3 protein pathogenesis. **A**. Flies expressing similar levels of SCA3 with a (**b**) pure CAG or an (**c**) interrupted CAA/CAG repeat show strikingly different degeneration. Flies with an interrupted CAA/CAG repeat show mild degeneration, with normal external eye morphology, compared to flies with a pure CAG repeat RNA. Normal fly in (**a**).**B**. Neuronal toxicity by life span analysis. Flies expressing the pure CAG repeat (pink line) have a strikingly shorter life span than flies expressing a toxic polyglutamine protein with an interrupted CAA/CAG repeat (blue line) at the same level (figure from Li et al. 2008)

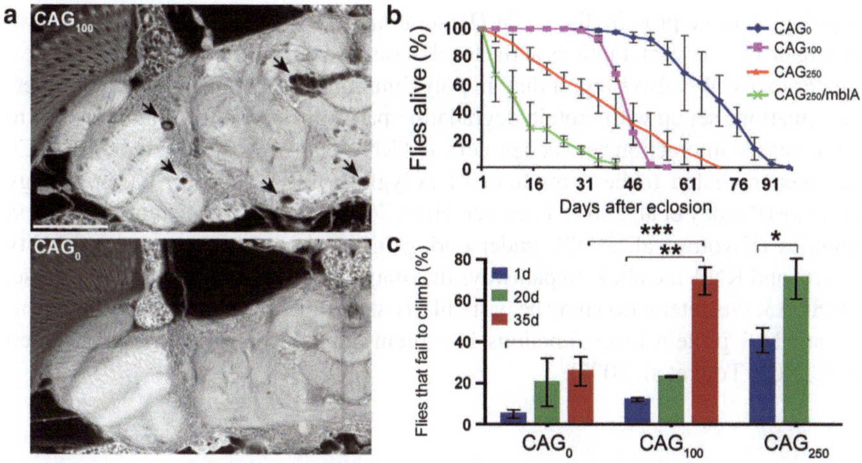

Fig. 4 Untranslated CAG repeats induce progressive neural dysfunction. **a.** Flies expressing untranslated CAG repeats show neuronal degeneration. Top, flies expressing a CAG repeat of 100 units showed progressive brain degeneration with vacuoles in the brain (arrows) at 35 days. Bottom, normal at 35 days. **b.** Expression of untranslated CAG repeats induces length-dependent, progressive neural dysfunction. Neurotoxicity by life span analysis. Flies expressing untranslated CAG repeats show length-dependent, reduced life span. **c.** Climbing ability with age. Flies with only the control protein have normal climbing defects with age (* $P<0.05$). Flies expressing a CAG repeat of 100 units had mild climbing defects at 20 days, which was worse at 35 days (*** $P<0.001$, ** $P<0.01$). Flies with repeats of 250 CAG units had moderate climbing defects at 1 day, which were strikingly worse by 20 days (* $P<0.05$; figure from Li et al. 2008)

We suggest that repeat expansions within RNAs would confer greater stability on the RNAs, indicating that, in the disease situation, any transcripts bearing the expanded repeat may accumulate aberrantly and confer deleterious effects that may contribute to disease. Moreover, it has become clear that over 70% of the human genome is bidirectionally transcribed, with transcripts being generated by both strands of any one gene (Mercer et al. 2009). Bi-directional transcripts bearing the expanded repeats are thought to occur in the CUG diseases myotonic dystrophy and SCA8, as well as other diseases like Fragile X (a CGG repeat expansion; Cho et al. 2005; Ladd et al. 2007; Moseley et al. 2006). Although research is only beginning on the implications of bi-directional transcription in these disease situations, research to date suggests a range of potential outcomes. For myotonic dystrophy, the expressions of both an expanded CTG in the DMPK gene and the anti-sense-expanded CAG transcript have been shown to be able to generate a double-stranded RNA that can become cleaved to small RNAs, resulting in local gene silencing (Cho et al. 2005). In the situation of SCA8, the anti-sense CAG transcript is thought to be able to be translated into a polyQ protein of pathogenic length, although this does not rule out effects occurring at the level of the RNA. Moreover, it is now clear that pre-expansion alleles of fragile-X cause fragile X tremor ataxia syndrome (FXTAS), which is due to the aberrant accumulation of the pre-mutation RNA (Hagerman and Hagerman 2004) and is modulated in the identical manner by molecular chaperones as polyglutamine protein disease in *Drosophila* models (Jin et al. 2003). Thus, we anticipate that new interactions at the level of disease transcript RNAs will become a greater focus. We also suggest that, in polyglutamine disease, there may be dangerous situations set up with protein degradation pathways becoming disrupted, due to the increased interaction between protein and RNA pathways. For example, HDAC6 has been revealed to be a modifier of polyglutamine disease in the autophagy pathway (Pandey et al. 2007); however, HDAC6 is a core component of RNA stress granules (Kwon et al. 2007), underscoring interactions between protein quality control and RNA metabolism pathways that may be critical to degenerative disease. For SCA3, this interaction may be particularly striking or sensitive to modulation, as the ataxin-3 protein itself functions in protein quality control pathways (Burnett et al. 2003; Todi et al. 2007).

4 Conclusions

Drosophila has proved a powerful system for revealing modifier pathways of neurodegeneration of striking relevance to human disease. *Drosophila* has revealed the critical importance of protein quality control pathways and now has revealed the importance of RNA pathways to polyglutamine disease. The continued use and application of fly genetics to such degenerative pathways, as well as to other disease models (Lessing and Bonini 2009), will prove of great value in providing mechanistic insight and defining potential pathways of greatest interest and effectiveness for therapeutic approaches.

Acknowledgments Research in the Bonini laboratory is supported by the NINDS, the NIA and the Howard Hughes Medical Institute. NMB is an Investigator of the Howard Hughes Medical Institute.

References

Ambros V (2004) The functions of animal microRNAs. Nature 431:350–355

Auluck PK, Chan HY, Trojanowski JQ, Lee VM, Bonini NM (2002) Chaperone suppression of alpha-synuclein toxicity in a Drosophila model for Parkinson's disease. Science 295:865–868

Bier E (2005) Drosophila, the golden bug, emerges as a tool for human genetics. Nature Rev Genet 6:9–23

Bilen J, Bonini NM (2005) Drosophila as a model for human neurodegenerative disease. Annu Rev Genet 39:153–171

Bilen J, Bonini NM (2007) Genome-wide screen for modifiers of ataxin-3 neurodegeneration in Drosophila. PLoS Genet 3:1950–1964

Bilen J, Liu N, Bonini NM (2006a) A new role for microRNA pathways: modulation of degeneration induced by pathogenic human disease proteins. Cell Cycle 5:2835–2838

Bilen J, Liu N, Burnett BG, Pittman RN, Bonini NM (2006b) MicroRNA pathways modulate polyglutamine-induced neurodegeneration. Mol Cell 24:157–163

Bonini NM (2002) Chaperoning brain degeneration. Proc Natl Acad Sci USA 99 Suppl 4:16407–16411

Brennecke J, Hipfner DR, Stark A, Russell RB, Cohen SM (2003) bantam encodes a developmentally regulated microRNA that controls cell proliferation and regulates the proapoptotic gene hid in Drosophila. Cell 113:25–36

Burnett B, Li F, Pittman RN (2003) The polyglutamine neurodegenerative protein ataxin-3 binds polyubiquitylated proteins and has ubiquitin protease activity. Human Mol Genet 12:3195–3205

Cho DH, Thienes CP, Mahoney SE, Analau E, Filippova GN, Tapscott SJ (2005) Antisense transcription and heterochromatin at the DM1 CTG repeats are constrained by CTCF. Mol Cell 20:483–489

Dedmon MM, Christodoulou J, Wilson MR, Dobson CM (2005) Heat shock protein 70 inhibits α-synuclein fibril formation via preferential binding to prefibrillar species. J Biol Chem 280:14733-14740

Hagerman PJ, Hagerman RJ (2004) The fragile-X premutation: a maturing perspective. Am J Human Genet 74:805–816

Hebert SS, Horre K, Nicolai L, Papadopoulou AS, Mandemakers W, Silahtaroglu AN, Kauppinen S, Delacourte A, De Strooper B (2008) Loss of microRNA cluster miR-29a/b-1 in sporadic Alzheimer's disease correlates with increased BACE1/β-secretase expression. Proc Natl Acad Sci USA 105:6415–6420

Jin P, Zarnescu DC, Zhang F, Pearson CE, Lucchesi JC, Moses K, Warren ST (2003) RNA-mediated neurodegeneration caused by the fragile X premutation rCGG repeats in Drosophila. Neuron 39:739–747

Kuyumcu-Martinez NM, Wang GS, Cooper TA (2007) Increased steady-state levels of CUGBP1 in myotonic dystrophy 1 are due to PKC-mediated hyperphosphorylation. Mol Cell 28:68–78

Kwon S, Zhang Y, Matthias P (2007) The deacetylase HDAC6 is a novel critical component of stress granules involved in the stress response. Genes Dev 21:3381–3394

Ladd PD, Smith LE, Rabaia NA, Moore JM, Georges SA, Hansen RS, Hagerman RJ, Tassone F, Tapscott SJ, Filippova GN (2007) An antisense transcript spanning the CGG repeat region of FMR1 is upregulated in premutation carriers but silenced in full mutation individuals. Hum Mol Genet 16:3174–3187

Lee VM, Trojanowski JQ (2006) Mechanisms of Parkinson's disease linked to pathological alpha-synuclein: new targets for drug discovery. Neuron 52:33–38

Lee YS, Nakahara K, Pham JW, Kim K, He Z, Sontheimer EJ, Carthew RW (2004) Distinct roles for Drosophila Dicer-1 and Dicer-2 in the siRNA/miRNA silencing pathways. Cell 117:69–81

Lessing D, Bonini NM (2009) Maintaining the brain: insight into human neurodegeneration from Drosophila mutants. Nature Rev Genet 10:359–370

Li LB, Yu Z, Teng X, Bonini NM (2008) RNA toxicity is a component of ataxin-3 degeneration in Drosophila. Nature 453:1107–1111

Marsh JL, Thompson LM (2006) Drosophila in the study of neurodegenerative disease. Neuron 52:169–178

Mercer TR, Dinger ME, Mattick JS (2009) Long non-coding RNAs: insights into functions. Nature Rev Genet 10:155–159

Miller JW, Urbinati CR, Teng-Umnuay P, Stenberg MG, Byrne BJ, Thornton CA, Swanson MS (2000) Recruitment of human muscleblind proteins to (CUG)(n) expansions associated with myotonic dystrophy. Embo J 19:4439–4448

Moseley ML, Zu T, Ikeda Y, Gao W, Mosemiller AK, Daughters RS, Chen G, Weatherspoon MR, Clark HB, Ebner TJ, Day JW, Ranum LP (2006) Bidirectional expression of CUG and CAG expansion transcripts and intranuclear polyglutamine inclusions in spinocerebellar ataxia type 8. Nature Genet 38:758–769

Orr HT, Zoghbi HY (2007) Trinucleotide repeat disorders. Annu Rev Neurosci 30:575–621

Pandey UB, Nie Z, Batlevi Y, McCray BA, Ritson GP, Nedelsky NB, Schwartz SL, DiProspero NA, Knight MA, Schuldiner O, Padmanabhan R, Hild M, Berry DL, Garza D, Hubbert CC,

Yao TP, Baehrecke EH, Taylor JP (2007) HDAC6 rescues neurodegeneration and provides an essential link between autophagy and the UPS. Nature 447:859–863

Pearson CE, Nichol Edamura K, Cleary JD (2005) Repeat instability: mechanisms of dynamic mutations. Nature Rev Genet 6:729–742

Ranum LP, Cooper TA (2006) RNA-mediated neuromuscular disorders. Annu Rev Neurosci 29:259–277

Saba R, Goodman CD, Huzarewich RL, Robertson C, Booth SA (2008) A miRNA signature of prion induced neurodegeneration. PLoS ONE 3:e3652

Schaefer A, O'Carroll D, Tan CL, Hillman D, Sugimori M, Llinas R, Greengard P (2007) Cerebellar neurodegeneration in the absence of microRNAs. J Exp Med 204:1553–1558

Selkoe DJ (2004) Cell biology of protein misfolding: the examples of Alzheimer's and Parkinson's diseases. Nature Cell Biol 6:1054–1061

Sobczak K, de Mezer M, Michlewski G, Krol J, Krzyzosiak WJ (2003) RNA structure of trinucleotide repeats associated with human neurological diseases. Nucleic Acids Res 31:5469–5482

Todi SV, Laco MN, Winborn BJ, Travis SM, Wen HM, Paulson HL (2007) Cellular turnover of the polyglutamine disease protein ataxin-3 is regulated by its catalytic activity. J Biol Chem 282:29348–29358

Warrick JM, Chan HY, Gray-Board GL, Chai Y, Paulson HL, Bonini NM (1999) Suppression of polyglutamine-mediated neurodegeneration in Drosophila by the molecular chaperone HSP70. Nature Genet 23:425–428

Wheeler TM, Thornton CA (2007) Myotonic dystrophy: RNA-mediated muscle disease. Curr Opin Neurol 20:572–576

Williams AJ, Paulson HL (2008) Polyglutamine neurodegeneration: protein misfolding revisited. Trends Neurosci 31:521–528

Wu YR, Wang CK, Chen CM, Hsu Y, Lin SJ, Lin YY, Fung HC, Chang KH, Lee-Chen GJ (2004) Analysis of heat-shock protein 70 gene polymorphisms and the risk of Parkinson's disease. Human Genet 114:236–241

Zhou Y, Gu G, Goodlett DR, Zhang T, Pan C, Montine TJ, Montine KS, Aebersold RH, Zhang J (2004) Analysis of α-synuclein-associated proteins by quantitative proteomics. J Biol Chem 279:39155–39164

Zuker M (2003) Mfold web server for nucleic acid folding and hybridization prediction. Nucl Acids Res 31:3406–3415

microRNAs in Sporadic Alzheimer's Disease and Related Dementias

Sébastien S. Hébert, Wim Mandemakers, Aikaterini S. Papadopoulou, and Bart DeStrooper

Abstract Recent studies have demonstrated that non-coding microRNAs (miRNAs), which function at the posttranscriptional level as a rheostat of the transcriptome and proteome, control a variety of neuronal functions as well as neuronal survival. Studies performed in humans support the idea that changes in miRNA expression profiles or target sequences could significantly contribute to the risk of major neurodegenerative diseases such as Alzheimer's disease (AD) and Parkinson's disease (PD). MiRNAs seem to participate directly in the regulation of expression of AD-related genes, including APP and BACE1/β-secretase, which are involved in the neurotoxic Aβ peptide production; the latter accumulates in the brains of AD patients. This observation is interesting, as gene dosage effects of the *APP* gene can cause genetic AD. In this regard, miRNA research appears to be particularly promising for the understanding of the very frequent and poorly understood sporadic forms of AD and probably other neurological disorders.

1 Introduction

The recently discovered gene family of microRNAs (miRNAs), small ~19-23-nt, evolutionary conserved non-coding RNAs, functions at the posttranscriptional level as a rheostat of the transcriptome and proteome of the cell (Ambros 2004). Similar to classical protein-coding genes, miRNA genes are embedded in the genome (at least 706 are found in humans; http://microrna.sanger.ac.uk) and are mostly transcribed by RNA polymerase II (Lee et al. 2004). Mechanistically, mature (fully processed) miRNAs modulate gene expression by binding with partial complementarity to the 3' untranslated region (3'UTR) of target messenger RNAs (mRNAs),

S.S. Hébert (✉)
Centre de Recherche du CHUQ (CHUL), Axe Neurosciences, Université Laval, Département de Biologie médicale, 2705 Boul. Laurier, Québec G1V 4G2, Local RC-9800, Québec, QcCanada
e-mail: Hébert@crchul.ulaval.ca

B. de Strooper and Y. Christen (eds.), *Macro Roles for MicroRNAs in the Life and Death of Neurons*, Research and Perspectives in Neurosciences,
DOI 10.1007/978-3-642-04298-0_10,

leading to their translational inhibition or degradation. The way in which miRNAs control translational repression remains under intense scrutiny but there is evidence involving the inhibition of translation initiation and/or deadenylation of target mRNA transcripts (Eulalio et al. 2008).

A particularly diverse and abundant source of miRNAs is the brain (Barad et al. 2004; Miska et al. 2004; Sempere et al. 2004), where they may be participating in various neuronal functions, including neurite outgrowth (Abdelmohsen et al. 2008), synapse formation (Schratt et al. 2006) and dentritogenesis (Fiore et al. 2009). Also known as the "master regulators" of gene expression, miRNAs are predicted to regulate as much as 25% of all protein-coding genes. A dysregulation of the miRNA network in neurons, which express a large number of genes, could have dramatic effects. Similarly, it is not surprising that genetic ablation (or reduction) of all miRNAs in the brain leads to progressive neurodegeneration in mammals and premature death (Davis et al. 2008; Kim et al. 2007; Schaefer et al. 2007). Whether neuronal loss is caused by the fine-tuning of a large array of genes or by the effects on a small number of key target genes remains unknown.

Given their impact on all major biological systems, miRNAs could contribute significantly to human diseases. This concept is already well documented in the cancer field, where the use of miRNAs as diagnostic and possibly therapeutic tools is increasingly acknowledged. A role for miRNAs in neurological disorders is slowly emerging as well. Sequence variants associated with Tourette's syndrome that affect miRNA binding in the 3′UTR of SLITRK1 (Abelson et al. 2005) and changes in miRNA expression profiles in schizophrenia (Perkins et al. 2007) and Down syndrome patients (Kuhn et al. 2008) are examples of miRNAs potentially involved in neurological disease.

2 Gene dosage effects in neurodegenerative disorders: a role for miRNAs?

In recent years, it has become well established that variations in the expression of genes can lead to the development of certain types of Alzheimer's disease (AD) and other neurological disorders. Indeed, *APP* gene duplications, as well as promoter mutations that increase APP expression, have been associated with familial AD, whereas an increase in BACE1/β-secretase protein has been linked to sporadic AD (Fukumoto et al. 2002; Hébert et al. 2008; Yang et al. 2003). BACE1 cleavage of APP is the rate-limiting step for amyloid-β (Aβ) peptide production in the brain, a factor that is believed to be crucial for the pathogenesis of AD (Haass 2004; Vassar 2001). Similarly, gene dosage effects of the α-synuclein (SNCA) gene have been associated with genetic Parkinson's disease (PD) (Singleton et al. 2003). Therefore, the study of the molecular mechanisms that regulate the expression of these and other genes is of great importance for a global understanding of the pathologies of AD and PD.

3 microRNAs in sporadic AD

Taking into account the growing evidence that the miRNA network is a major regulator of gene expression, it is reasonable to speculate that loss of the fine tuning of this network could be relevant for sporadic AD (Hébert and De Strooper 2007). Moreover, knowing that the greatest risk factor for an individual to develop AD is age, an association between aging and such loss of function cannot be ruled out.

To gain insight into the role of miRNAs in AD development, our group performed extensive microarray analyses of the brains of sporadic AD patients and appropriate age-matched controls (Fig. 1A), and we found that several miRNAs were significantly altered in AD (Fig. 1B; Hébert et al. 2008). Surprisingly, in our sample set, several (7 of 16 significantly altered) miRNAs potentially involved in the regulation of BACE1 and APP expression appeared to be decreased in diseased brain (Fig. 1B and C). Mechanistically, decreased miRNA levels could lead to increased protein levels (while mRNA levels remain unchanged).

We corroborated the hypothesis that specific miRNAs regulate BACE1 (e.g., miR-29a and miR-29b-1) and APP (e.g., miR-106b) in various in vitro and cell culture paradigms (Hébert et al. 2008, 2009 Hébert). The regulation of human APP expression in vitro by miRNAs of the miR-106b family (i.e., miR-106b, miR-106a, miR-20a and miR-17-5p) was independently confirmed (Patel et al. 2008).

Analysis of brain extracts of sporadic AD patients revealed that roughly 30% (11/34) displayed increased BACE1 protein levels (Hébert et al. 2008), which is in agreement with previous findings (Fukumoto et al. 2002; Yang et al. 2003). The observation that BACE1 mRNA levels remained quite stable overall in AD patients (Hébert et al. 2008) suggests that BACE1 is, at least partly, regulated at the posttranscriptional level. A statistically significant (and AD dementia-specific) correlation between BACE1 and changes in miR-29a/miR-29b-1 expression was observed in AD patients (Hébert et al. 2008). Although a significant decrease in miR-106b expression was observed in all AD cases, a correlation with APP expression could not be further established, as APP levels, even in control brains, varied greatly (Hébert et al. 2008, 2009 Hébert). Further work is therefore needed to firmly establish the link between miR-106b and APP expression in vivo. Interestingly, correlations between expression of these miRNAs and BACE1 and APP expression were found in non-pathological conditions in the mouse brain during embryonic development and in primary cell cultures (Hébert et al. 2008). In addition, evidence for a potential causal relationship between miR-29a/miR-29b-1 expression, BACE1 and Aβ generation were obtained in a cell culture model (Hébert et al. 2008). Overall, these results provide strong circumstantial attestations that a loss-of-function of specific miRNAs could contribute to the increased Aβ production and subsequently the risk for sporadic AD. This finding is of great importance since such a mechanism (loss-of-function of specific miRNAs) could augment regulation of BACE1 (and perhaps APP) expression in the aging and compromised brain. In addition, these observations also provide an interesting molecular link between sporadic AD and the amyloid cascade.

a

Samples	Age	Apoe	Tau staging	Amyloid staging	Clinical diagnostic	Pathology	RNA 260/280 ratio	RIN
C1	82	E3/E3	S3	S1,S0	Ctrl	control	2,10	6,7
C2	69	E2/E3	S4	S0,	Ctrl	control	2,06	6,3
C3	51	E2/E3	S3	S0,S0	Ctrl	control	2,13	6,5
C4	77	E2/E3	S3	S5,S0	Ctrl	control	2,12	5,7
C5	65	E3/E4	S0	S0,S0	Ctrl	control	2,10	6,6
AD1	74	E4/E4	S10	S9,S8	Pr Alz	Def Alz	2,09	7,5
AD2	83	E3/E3	S10	S7,	Pr Alz	Def Alz	2,08	6,6
AD3	69	E2/E4	S10	S8,S7	pr semantic dementia	Def Alz	2,11	5,9
AD4	64	E3/E4	S7	S7,S0	po Alz, po LBD, po Vasc	Def Alz	2,09	6,3
AD5	84	E2/E3	S9	S6,S5	dementia	mixte dementia	2,09	7,2

b

BioArray ID	miR Name	p-values	% Present (All Samples)	% Present (Controls)	% Present (AD)	Fold Change
BM10516	hsa_miR_210	1,61556E-05	70	100	40	0,66
BM10181	hsa_miR_181c	0,00183884	100	100	100	0,71
BM10235	**hsa_miR_15a**	0,003063763	100	100	100	0,65
BM10022	**hsa_miR_9**	0,005306359	100	100	100	0,77
BM10203	hsa_miR_22	0,014002865	100	100	100	0,83
BM10537	**hsa_miR_101**	0,016000385	60	80	40	0,69
BM10354	hsa_miR_197	0,018172544	100	100	100	1,25
BM10237	hsa_miR_511	0,020522018	80	80	80	1,21
BM10103	**hsa_miR_29b**	0,027514137	100	100	100	0,6
BM10517	**hsa_let_7i**	0,028267592	100	100	100	0,88
BM10629	**hsa_miR_19b**	0,033686716	100	100	100	0,64
BM10214	hsa_miR_320	0,035521612	100	100	100	1,15
BM10067	**hsa_miR_106b**	0,03643293	90	100	80	0,71
BM11070	hsa_miR_26b	0,039267772	100	100	100	0,79
BM10149	hsa_miR_363	0,044045684	90	100	80	0,76
BM10318	hsa_miR_93	0,046688633	100	100	100	0,83

c

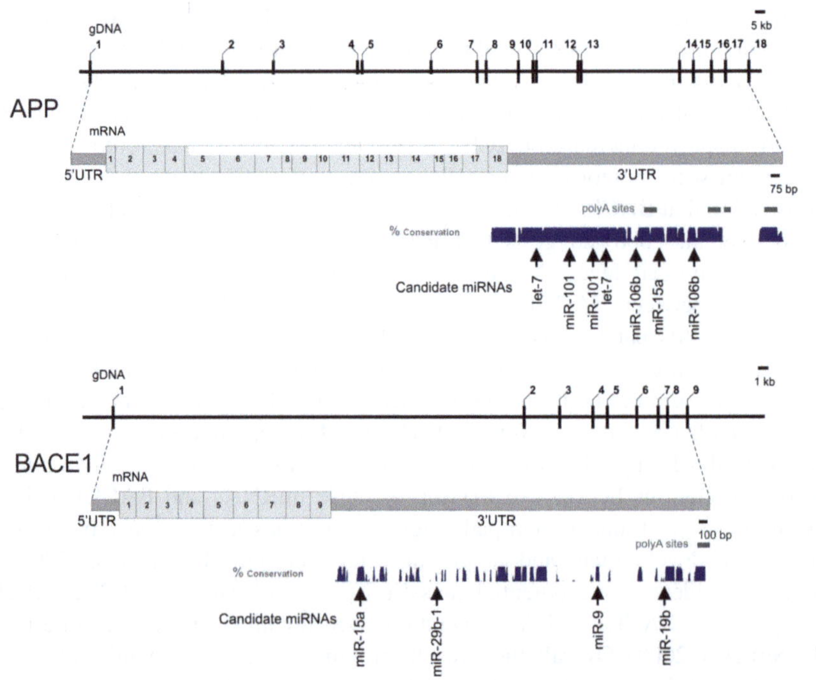

Fig. 1 Changes in miRNA expression profiles in sporadic AD brain. (**A**) Neuropathological, biochemical and clinical assessment of patient brain samples (temporal cortex) used for miRNA profiling studies (Hébert et al. 2008). Here, five controls (C) and five aged-matched AD cases were profiled in parallel. Assessment of total RNA quality obtained from postmortem tissues is indicated as the RNA 260:280 ratio as well as the RNA integrity number (RIN). Specific details can be found in the Hébert et al. (2008). (**B**) Summary of statistically significantly changed

Other groups have likewise reported changes in miRNA expression profiles in the AD brain (Cogswell et al. 2008; Lukiw 2007; Wang et al. 2008). While little or no obvious overlap was observed with our profiling results, Wang and colleagues (2008) identified additional miRNAs, such as miR-107, to be altered in diseased brain and potentially involved in BACE1 expression regulation. In this study, decreased miR-107 expression was observed at a clinical stage prior to "pure" AD dementia. Although definite conclusions cannot be drawn due to great variances in the profiling data from AD patients (Hébert and De Strooper 2009), it remains safe to anticipate that diseased or degenerative brain (or neurons) harbors changes in miRNA expression profiles. Whether these changes are specific to a particular disease or disease state remains to be confirmed.

Neurofibrillary tangles constitute the second main neuropathological hallmark of AD brain. These tangles are formed by hyperphosphorylation of the microtubule-associated protein tau, causing it to aggregate to a toxic, insoluble form (Delacourte and Buee 2000). Interestingly, a role for miRNA-regulated pathways in tau metabolism and toxicity is emerging. Studies by Bilen and colleagues (2006) showed a striking enhancement of tau toxicity in Drosophila cells when miRNA maturation was suppressed. More recently, a neuronal miRNA, miR-124, was shown to modulate BAG2 expression, a co-chaperone potentially involved in tau degradation and aggregation (Carrettiero et al. 2009). An independent study reports an increase in miR-124 expression in AD (Lukiw 2007). Given that several candidate miRNA binding sites exist in the 3'UTR of tau mRNA (www.targetscan.org), as well as of several enzymes involved in tau phosphorylation (e.g., MAPK1 and PPP2CA), it is tempting to speculate that changes in miRNA-regulated pathways could have an effect, either directly or indirectly, on tau physiology and pathophysiology. Finally, miRNAs could also be involved in more precise, perhaps even secondary (i.e., downstream of Aβ toxicity), aspects of AD neuropathology. For instance, work by Lukiw and colleagues (2008) suggests a role for miR-146a in the inflammation response in AD brain. These observations raise questions about the cause-effect relationship of miRNA dysregulation in AD development.

In conclusion, it is becoming increasingly apparent that loss of fine genetic regulation by the miRNA network, in combination with the increased Aβ load in the aging neurons, could provide a "multiple hit" scenario for the severe neurodegeneration observed in the AD brain.

Fig. 1 (Continued) miRNAs in AD brain. In total, 328 (167 above background levels) human miRNAs were analyzed. Most miRNAs seemed downregulated in diseased brains. Here, miRNA expression profiling and quantification were performed using the mirVana miRNA Probe Set V2 (Ambion, Austin, TX, USA). The bioarray miRNA identification number and the percentage of single miRNAs found in each group (n=5) are shown. For detailed statistical analysis, refer to Hébert et al. (2008). Of 16 significantly altered miRNAs in AD, seven are predicted to target APP or BACE1 genes (in bold). **(C)** Schematic representation of human APP and BACE1 genes with candidate (based on bioinformatics) miRNA binding sites in the 3'UTRs. The genomic (gDNA) and mRNA structures are shown (not to scale). The number of exons, the percentage of conservation in the 3'UTR, and the putative poly A sites are indicated

4 Clinical relevance for miRNAs in neurodegenerative disease

Although it is clear that miRNAs are involved in the regulation of genes causally linked to neurodegenerative disorders, it will be difficult to provide definitive proof for this concept at the clinical level in the near future. However, the very recent identification of patient-specific genetic mutations in miRNA-binding sites in the 3'UTR of APP and BACE1 provides initial evidence for this hypothesis (Bettens et al. 2009). The challenge now is to establish a causal and functional relationship between these miRNAs and APP and BACE1 expression with the disease in animal models and, ultimately, in patients. However, as already noted, the functional relevance for AD remains to be explored.

Interestingly, the study by Bettens and colleagues (2009) also identified AD patient-specific polymorphisms in the genomic region of the miR-29a/b-1 cluster. Functional polymorphisms in miRNA genes have previously been associated with cancer (Calin and Croce 2006; Xu et al. 2008).

Polymorphisms in miRNA sequences or target genes may play a role in other neurodegenerative diseases as well. For instance, Wang and colleagues (2008) showed that a single nucleotide polymorphism (SNP) located in the 3′UTR of the fibroblast growth factor 20 (FGF20) gene confers risk for developing PD, possibly by loss of miR-433 binding. The FGF20 protein induces α-synuclein expression in cell cultures and, thus, lack of FGF20 suppression could account for concomitant increases in α-synuclein expression in affected patients. In addition, Rademakers and colleagues (2008) showed that increased binding of miR-659 to the 3′UTR of the progranulin (GRN) gene provides an important risk for TAR DNA-binding protein 43 (TDP43)-positive frontotemporal dementia. Loss of function of the GRN gene is causally linked to PD.

5 Conclusion and Perspectives

Hitherto, miRNA research has mainly focused on cancer. The emergence of miRNAs in the field of neurodegeneration could shed light on the regulation of the genes and, subsequently, cellular pathways potentially involved in neurodegenerative disorders. The current literature points out that misregulation of miRNAs could be a major cause of sporadic AD and probably PD.

References

Abdelmohsen K, Srikantan S, Kuwano Y, Gorospe M (2008) miR-519 reduces cell proliferation by lowering RNA-binding protein HuR levels. Proc Natl Acad Sci USA 105:20297–20302

Abelson JF, Kwan KY, O'Roak BJ, Baek DY, Stillman AA, Morgan TM, Mathews CA, Pauls DL, Rasin MR, Gunel M, Davis NR, Ercan-Sencicek AG, Guez DH, Spertus JA, Leckman JF,

Dure LSt, Kurlan R, Singer HS, Gilbert DL, Farhi A, Louvi A, Lifton RP, Sestan N, State MW (2005) Sequence variants in SLITRK1 are associated with Tourette's syndrome. Science 310:317–320

Ambros V (2004) The functions of animal microRNAs. Nature 431:350–355

Barad O, Meiri E, Avniel A, Aharonov R, Barzilai A, Bentwich I, Einav U, Gilad S, Hurban P, Karov Y, Lobenhofer EK, Sharon E, Shiboleth YM, Shtutman M, Bentwich Z, Einat P (2004) MicroRNA expression detected by oligonucleotide microarrays: system establishment and expression profiling in human tissues. Genome Res 14:2486–2494

Bettens K, Brouwers N, Engelborghs S, van Miegroet H, De Deyn P, Theuns J, Sleegers K, Van Broeckhoven C (2009) APP and BACE1 miRNA genetic variability has no major role in risk for Alzheimer's disease. Human Mutat 30:1207–1213

Bilen J, Liu N, Burnett BG, Pittman RN, Bonini NM (2006) MicroRNA pathways modulate polyglutamine-induced neurodegeneration. Mol Cell 24:157–163

Calin GA, Croce CM (2006) MicroRNA-cancer connection: the beginning of a new tale. Cancer Res 66:7390–7394

Carrettiero DC, Hernandez I, Neveu P, Papagiannakopoulos T, Kosik KS (2009) The cochaperone BAG2 sweeps paired helical filament- insoluble tau from the microtubule. J Neurosci 29:2151–2161

Cogswell JP, Ward J, Taylor IA, Waters M, Shi Y, Cannon B, Kelnar K, Kemppainen J, Brown D, Chen C, Prinjha RK, Richardson JC, Saunders AM, Roses AD, Richards CA (2008) Identification of miRNA changes in Alzheimer's disease brain and CSF yields putative biomarkers and insights into disease pathways. J Alzheimers Dis 14:27–41

Davis TH, Cuellar TL, Koch SM, Barker AJ, Harfe BD, McManus MT, Ullian EM (2008) Conditional loss of Dicer disrupts cellular and tissue morphogenesis in the cortex and hippocampus. J Neurosci 28:4322–4330

Delacourte A, Buee L (2000. Tau pathology: a marker of neurodegenerative disorders. Curr Opin Neurol 13:371–376

Eulalio A, Huntzinger E, Izaurralde E (2008) Getting to the root of miRNA-mediated gene silencing. Cell 132:9–14

Fiore R, Khudayberdiev S, Christensen M, Siegel G, Flavell SW, Kim TK, Greenberg ME, Schratt G (2009) Mef2-mediated transcription of the miR379-410 cluster regulates activity-dependent dendritogenesis by fine-tuning Pumilio2 protein levels. EMBO J 28:697–710

Fukumoto H, Cheung BS, Hyman BT, Irizarry MC (2002) ß-secretase protein and activity are increased in the neocortex in Alzheimer disease. Arch Neurol 59:1381–1389

Haass C (2004) Take five–BACE and the r-secretase quartet conduct Alzheimer's amyloid ß-peptide generation. EMBO J 23:483–488

Hébert SS, De Strooper B (2007) Molecular biology. miRNAs in neurodegeneration. Science 317:1179–1180

Hébert SS, De Strooper B (2009) Alterations of the microRNA network cause neurodegenerative disease. Trends Neurosci 32:199–206

Hébert SS, Horre K, Nicolai L, Papadopoulou AS, Mandemakers W, Silahtaroglu AN, Kauppinen S, Delacourte A, De Strooper B (2008) Loss of microRNA cluster miR-29a/b-1 in sporadic Alzheimer's disease correlates with increased BACE1/ß-secretase expression. Proc Natl Acad Sci USA 105:6415–6420.

Hébert SS, Horre K, Nicolai L, Bergmans B, Papadopoulou AS, Delacourte A, De Strooper B (2009) MicroRNA regulation of Alzheimer's amyloid precursor protein expression. Neurobiol Dis 33:422–428

Kim J, Inoue K, Ishii J, Vanti WB, Voronov SV, Murchison E, Hannon G, Abeliovich A (2007) A MicroRNA feedback circuit in midbrain dopamine neurons. Science 317:1220–1224

Kuhn DE, Nuovo GJ, Martin MM, Malana GE, Pleister AP, Jiang J, Schmittgen TD, Terry AV, Jr., Gardiner K, Head E, Feldman DS, Elton TS (2008) Human chromosome 21-derived miRNAs are overexpressed in down syndrome brains and hearts. Biochem Biophys Res Commun 370:473–477

Lee Y, Kim M, Han J, Yeom KH, Lee S, Baek SH, Kim VN (2004) MicroRNA genes are transcribed by RNA polymerase II. EMBO J 23:4051–4060
Lukiw WJ (2007) Micro-RNA speciation in fetal, adult and Alzheimer's disease hippocampus. Neuroreport 18:297–300
Lukiw WJ, Zhao Y, Cui JG (2008) An NF-?B-sensitive micro RNA-146a-mediated inflammatory circuit in Alzheimer disease and in stressed human brain cells. J Biol Chem 283:31315–31322
Miska EA, Alvarez-Saavedra E, Townsend M, Yoshii A, Sestan N, Rakic P, Constantine-Paton M, Horvitz HR (2004) Microarray analysis of microRNA expression in the developing mammalian brain. Genome Biol 5:R68
Patel N, Hoang D, Miller N, Ansaloni S, Huang Q, Rogers JT, Lee JC, Saunders AJ (2008) MicroRNAs can regulate human APP levels. Mol Neurodegener 3:10
Perkins DO, Jeffries CD, Jarskog LF, Thomson JM, Woods K, Newman MA, Parker JS, Jin J, Hammond SM (2007) microRNA expression in the prefrontal cortex of individuals with schizophrenia and schizoaffective disorder. Genome Biol 8:R27
Rademakers R, Eriksen JL, Baker M, Robinson T, Ahmed Z, Lincoln SJ, Finch N, Rutherford NJ, Crook RJ, Josephs KA, Boeve BF, Knopman DS, Petersen RC, Parisi JE, Caselli RJ, Wszolek ZK, Uitti RJ, Feldman H, Hutton ML, Mackenzie IR, Graff-Radford NR, Dickson DW (2008) Common variation in the miR-659 binding-site of GRN is a major risk factor for TDP43-positive frontotemporal dementia. Human Mol Genet 17:3631–3642
Schaefer A, O'Carroll D, Tan CL, Hillman D, Sugimori M, Llinas R, Greengard P (2007) Cerebellar neurodegeneration in the absence of microRNAs. J Exp Med 204:1553–1558
Schratt GM, Tuebing F, Nigh EA, Kane CG, Sabatini ME, Kiebler M, Greenberg ME (2006) A brain-specific microRNA regulates dendritic spine development. Nature 439:283–289
Sempere LF, Freemantle S, Pitha-Rowe I, Moss E, Dmitrovsky E, Ambros V (2004) Expression profiling of mammalian microRNAs uncovers a subset of brain-expressed microRNAs with possible roles in murine and human neuronal differentiation. Genome Biol 5:R13
Singleton AB, Farrer M, Johnson J, Singleton A, Hague S, Kachergus J, Hulihan M, Peuralinna T, Dutra A, Nussbaum R, Lincoln S, Crawley A, Hanson M, Maraganore D, Adler C, Cookson MR, Muenter M, Baptista M, Miller D, Blancato J, Hardy J, Gwinn-Hardy K (2003) a-Synuclein locus triplication causes Parkinson's disease. Science 302:841
Vassar R (2001) The ß-secretase, BACE: a prime drug target for Alzheimer's disease. J Mol Neurosci 17:157–170
Wang G, van der Walt JM, Mayhew G, Li YJ, Zuchner S, Scott WK, Martin ER, Vance JM (2008) Variation in the miRNA-433 binding site of FGF20 confers risk for Parkinson disease by overexpression of a-synuclein. Am J Human Genet 82:283–289
Wang WX, Rajeev BW, Stromberg AJ, Ren N, Tang G, Huang Q, Rigoutsos I, Nelson PT (2008) The expression of microRNA miR-107 decreases early in Alzheimer's disease and may accelerate disease progression through regulation of ß-site amyloid precursor protein-cleaving enzyme 1. J Neurosci 28:1213–1223
Xu T, Zhu Y, Wei QK, Yuan Y, Zhou F, Ge YY, Yang JR, Su H, Zhuang SM (2008) A functional polymorphism in the miR-146a gene is associated with the risk for hepatocellular carcinoma. Carcinogenesis 29:2126–2131
Yang LB, Lindholm K, Yan R, Citron M, Xia W, Yang XL, Beach T, Sue L, Wong P, Price D, Li R, Shen Y (2003) Elevated ß-secretase expression and enzymatic activity detected in sporadic Alzheimer disease. Nature Med 9:3–4

microRNA Dysregulation in Psychiatric Disorders

Bin Xu, Joseph A. Gogos, and Maria Karayiorgou

Abstract MicroRNAs (miRNAs) are a class of small, non-coding RNAs, about 22 nucleotides long that are important components of a highly conserved gene regulatory mechanism. Although abnormalities in miRNA-mediated gene regulation have been observed in a variety of human diseases, the study of miRNAs in psychiatric disorders and in normal neuronal development is still in its infancy. Individuals with 22q11.2 microdeletions are at high risk to develop schizophrenia. Here, we summarize recent findings from our laboratory using a mouse model of the human 22q11.2 locus, which led to the discovery of a previously unknown alteration in the biogenesis of miRNAs emerging as a result of the 22q11.2 microdeletion. We review available evidence indicating that the abnormal miRNA biogenesis emerges due to haploinsufficiency of the *Dgcr8* gene, which encodes for a RNA binding moiety of the "microprocessor" complex and contributes to the behavioral and neuronal deficits associated with the 22q11.2 microdeletion. To help interpret those findings in the context of the existing knowledge, we also outline our current knowledge about miRNA biogenesis and potential mode of action and summarize a number of current studies on the role of miRNAs in the developing and adult central nervous system, as well as on the emerging connections between miRNAs and neuropsychiatric diseases.

1 Introduction

MicroRNAs (miRNAs) are a class of small, non-coding RNAs, about 22 nucleotides (nt) long. Since the first miRNA, lin-4, was identified in the nematode *C. elegans*, a large amount of experimental evidence indicates miRNAs are

M. Karayiorgou (✉)
Department of Psychiatry, Columbia University, New York, NY USA
e-mail: mk2758@columbia.edu

B. de Strooper and Y. Christen (eds.), *Macro Roles for MicroRNAs in the Life and Death of Neurons*, Research and Perspectives in Neurosciences,
DOI 10.1007/978-3-642-04298-0_11,

important components of a highly conserved gene regulatory mechanism. Abnormalities in miRNA-mediated gene regulation have been observed in a variety of human diseases. The study of miRNAs in psychiatric and cognitive disorders, as well as in normal neuronal development, is still in its infancy but promises to provide important insights in the next few years. In this review, we will first outline our current knowledge about miRNA biogenesis and its potential mode of action. We will then summarize a number of current studies on the role of miRNAs in the developing and adult central nervous system (CNS). Finally, we will discuss the emerging connections between miRNAs and neuropsychiatric diseases (Fig. 1).

2 miRNA Biogenesis

2.1 miRNA biogenesis and RNAinduced silencing complex (RISC) pathway

Similar to mRNAs, most of the miRNAs are transcribed by an RNA polymerase II as long primary transcripts (pri-miRNAs) that contain 5'Cap, internal core hairpin structures and 3' poly(A) tails. The only exceptions are the miRNAs located in Alu-repetitive elements, which are transcribed by RNA polymerase III (Borchert et al. 2006). The hairpin structures of pri-miRNAs are cleaved in the nucleus by the microprocessor (a complex containing type-III RNase Drosha and its partner protein Dgcr8) into stem-loop precursor miRNAs about 70 nt long (pre-miRNAs). Pre-miRNAs are then exported to the cytoplasm by a process facilitated by the transport factor Exportin-5. In the cytoplasm, pre-miRNAs are further cleaved by another type-III RNase, Dicer. The resulting mature miRNA duplexes are processed by Dicer and several other RNA binding proteins, such as Ago2, PACT and TRBP, and one of the strands is then incorporated in the RISC. The miRNA-associated RISC binds to the target mRNA and inhibits its translation or cleaves the target transcript (Kim 2005).

miRNAs can also be produced by several non-canonical pathways. In one such pathway, the early processing step is accomplished by the combined action of the spliceosome and a debranching enzyme that yields a short hairpin, which is ready for further processing by Dicer. These non-canonical miRNAs have been termed mirtrons (Okamura et al. 2007; Ruby et al. 2007). In another pathway, short hairpin RNAs (shRNAs) are processed by unknown nucleases into pre-miRNAs, which are further processed into miRNAs by Dicer. miRNAs that are derived in this way have been termed endogenous shRNA-derived miRNAs (Babiarz et al. 2008).

2.2 Regulation of miRNA expression and maturation

Regulation of miRNA gene transcription in organisms is under both spatial and temporal control. In particular, brain miRNA gene transcription can be controlled

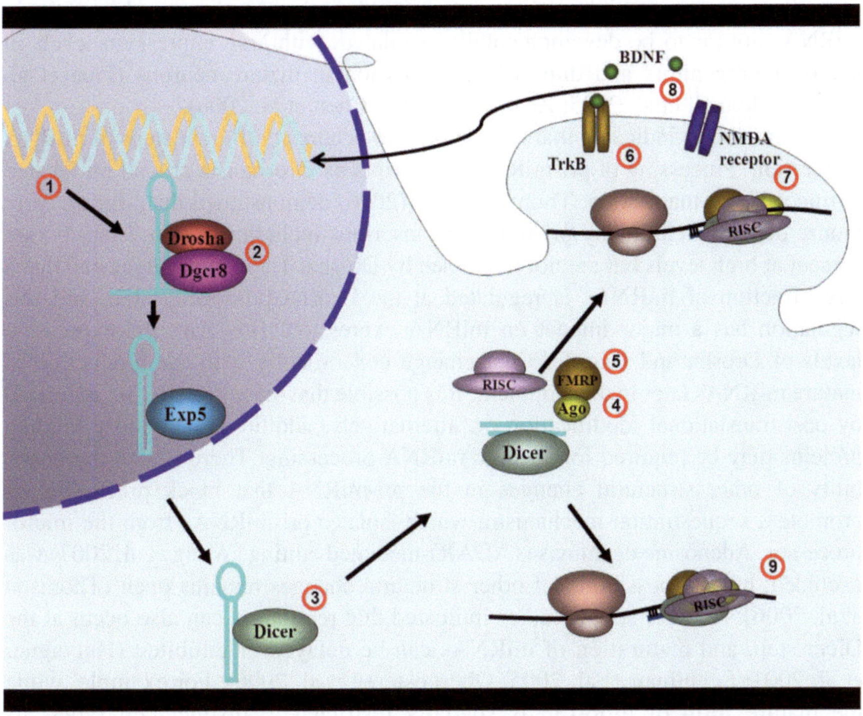

Fig. 1 miRNAs, neuronal function and neuropsychiatric disorders. (1) A number of miRNA gene expression profiling assays have identified alterations of miRNA gene expression in samples of patients with neuropsychiatric disorders. (2) *Dgcr8*$^{+/-}$ mice show smaller dendritic spines, reduced dendritic complexity, impaired working memory-based performance and sensorimotor gating function. The human orthologue, *DGCR8*, is located within the 22q11.2 locus. Microdeletions of this locus, resulting in cognitive deficits, are associated with a dramatic increase in the risk for schizophrenia and account for 1–2% of non-familial cases of this disease. (3) Conditional knockout of *Dicer* disrupts cellular and tissue morphogenesis and results in progressive neuron cell death in the forebrain and cerebellum. (4) *Ago-1* knockout mice show developmental defects with prominent malformation of the nervous system. *Ago-2*-null mice show severe neural tube defects. (5) FMRP hypermethylation and inactivation have been associated with fragile X mental retardation syndrome. (6) Several activity-dependent neuronal functions are modulated by miRNAs. For example, dendritic spine morphology is controlled by the inhibitory effect of miR-134 on Limk1 mRNA translation, which is, in turn, regulated by the synaptic activity-dependent BDNF induction. (7) The NMDA receptor transmission function is regulated by miR-219 through its target CamKII gamma. (8) Neuronal activity and BDNF release regulate the transcription of microRNAs, such as the mir379-410 cluster, which in turn regulates dendritogenesis. (9) Several candidate genes of neuropsychiatric disorders are potential targets of specific microRNAs. For example, Slit and Trk-like1 (*SLITRK1*), a candidate Tourette's syndrome gene, is a potential target of miR-189; MeCP2, the Rett syndrome gene, is a potential target of miR-132; BACE1, a candidate Alzheimer's disease gene, is a potential target of miR-107

by various pol-II-associated regulatory factors so that specific sets of miRNAs can be expressed during development, as well as in certain regions or cell types. For instance, miR-124 is expressed predominantly in the CNS, whereas miR-26 and miR-29 are present mainly in astrocytes. On the other hand, a number of brain miRNAs appear to be developmentally regulated, with high expression levels in neuronal progenitors and diminished expression in mature neurons (Fazi et al. 2005; O'Donnell et al. 2005; Zhao et al. 2005; Chen et al. 2006).

Recent studies indicated that regulation also occurs at different steps of miRNA maturation. Processing of pri-miRNA transcripts by Drosha is a crucial initial step during miRNA maturation. Thomson et al. (2006) demonstrated that, during early mouse development, many pri-miRNA transcripts, including the let-7 family, are present at high levels but are not processed by Drosha. Their study suggested that a large fraction of miRNAs is regulated at the Drosha processing step, and this regulation has a major impact on miRNA expression. However, the expression levels of Drosha and Dgcr8 do not change concordantly with the production of mature miRNAs later in development. It is possible that these proteins are regulated by post-translational modification or, alternatively, additional regulatory binding proteins may be required for specific miRNA processing. There is also the possibility of other structural changes in the pri-miRNA that block processing or promote a sequestration mechanism, which isolates pri-miRNAs from the microprocessor. Adenosine deaminase (ADAR)-mediated editing (Wang et al. 2007) was excluded, but the possibility of other structural changes remains open (Thomson et al. 2006). Another set of studies indicated that regulation can also occur at the Dicer step, and maturation of miRNAs can be delayed or inhibited (Hutvagner et al. 2001; Schulman et al. 2005; Obernosterer et al. 2006). For example, while the mature form of miR-138 is spatially restricted to distinct cell types, its precursor, pre-miR-138-2, is ubiquitously expressed in all tissues analyzed, which suggests that Dicer-mediated cleavage of this pre-miRNA is restricted to certain tissues and cell types. Thus, differential processing of pre-miRNAs might be an alternative mechanism to control miRNA function (Obernosterer et al. 2006).

3 Mode of action

The major and most intensively studied role of miRNAs is to control the translation and stability of their corresponding target mRNAs. In addition, several other potential regulatory roles have been identified or proposed.

3.1 Target control

In animals, it is now generally accepted that a "seed" region of a miRNA (2–7 nucleotides from its 5′ end) is important for the miRNA to recognize its target mRNAs. However, the accessibility and efficiency of target binding are likely dependent on many surrounding factors, such as the secondary structure of the target mRNA, the sequence context near the binding site, and the RNA binding protein complexes associated with the target mRNA and the miRNA. In most of the studied examples, miRNAs repress the target mRNA stability or translation, although activation of translation by miRNAs has also been reported recently (Vasudevan et al. 2007). The detailed mechanism by which miRNAs control stability or translation of mRNA remains controversial (Pillai et al. 2004; Humphreys et al. 2005; Maroney et al. 2006; Petersen et al. 2006; Behm-Ansmant et al. 2006; Eulalio et al. 2007). In addition, two models have been proposed to account for miRNA-target relationship, the "switch model" and the "tuning model" (Bartel and Chen, 2004).

Switch model: In the switch model, the translation of the targets is turned off or reduced to inconsequential levels by their miRNAs. A relevant extreme situation is the mutually exclusive expression of miRNAs and their targets in the same cells. For example, two brain-specific miRNAs (miR-124a and miR-9) have been implicated in the decision of a mouse neural precursor to adopt a neuronal or glial fate. Both gain-of-function and loss-of-function approaches have shown that miR-124a and miR-9 can shift the proportion of neurons, as compared with glia, during neural lineage differentiation (Krichevsky et al. 2006).

Tuning model: In contrast to the switch model, "tuning miRNAs" are co-expressed with their targets, set a stringent threshold for translational activity of the target mRNAs, and also help reduce stochastic cell-to-cell variability during developmental fate decisions. Such miRNAs act as a rheostat rather than binary off-switches to dampen protein output to a more optimal level and ensure uniform expression within each cell type (Cohen et al. 2006). Recently, quantitative mass spectrometry was used to measure the response of thousands of proteins after introducing or deleting specific miRNAs into cultured cells. Hundreds of genes containing seed-matched sites were directly repressed, albeit each to a modest degree, by individual miRNAs. Overall, consistent with the tuning model, the impact of miRNAs on the proteome indicated that, for most interactions, miRNAs act to make fine-scale adjustments to protein output (Baek et al. 2008; Selbach et al. 2008).

3.2 *Regulatory network control*

Individual organisms can have hundreds of miRNAs that have hundreds of targets, and therefore it is unlikely that we will be able to uniformly estimate the biological consequences of all potential interactions. Nevertheless, recent studies indicate that there are two general distinct, but not mutually exclusive, modes of miRNA action at the intracellular level: 1) targeting the expression of gene batteries (i.e., sets of functionally related effector genes that represent outputs of genetic regulatory networks) and 2) targeting the expression of critical regulators, which in turn control certain cellular functions (Makeyev and Maniatis 2008).

4 Characteristics of miRNA Action

miRNAs share many characteristics with conventional transcription factors (TFs) (Hobert 2008). For example, both can bind to discrete cis-regulatory elements and control many target genes, both use cooperativity as a mechanism for reading out combinatorial expression patterns, both exploit binding site accessibility as an additional layer of gene regulatory control and finally, both miRNAs and TFs recognize their target sequence in an imperfect match fashion. However, it is important to note that miRNAs present some unique features that render them more flexible in controlling specific aspects of terminal differentiation programs of individual cell types. For example, miRNA action can be spatially compartmentalized within a cell (i.e., in dendrites of neurons) to alter gene expression locally. Speed and reversibility are other distinguishing features of miRNA-mediated gene regulation that may result in a more specialized regulatory control of miRNAs over temporal gene regulation.

Two rapidly progressing areas of study of miRNA function in the nervous system concern their roles in nervous system development and in activity-dependent function and plasticity. A remarkable feature of the nervous system is that it contains many different types of neurons that have distinct dendritic architecture, different axonal targets and other functional specific characteristics, such as receptors and neurotransmitters on the cell surfaces. The impact of miRNAs upon these aspects in both the developing and mature nervous system has been assessed using a variety of strategies in various model organisms and human samples. We will summarize some of the findings from both model organisms and human samples that are relevant to the role of miRNAs in psychiatric diseases.

4.1 The impact of disrupting miRNA biogenesis and RISC pathways

Disruption of the Zebrafish *Dicer* gene and the ensuing suppression of miRNA biogenesis have provided the first evidence that miRNAs are necessary for the normal development of the nervous system (Giraldez et al. 2005). *Dicer* mutants showed severe defects in neural tube morphogenesis, arising from abnormal neuronal differentiation rather than deficits in early patterning or specification. In mammals, the role of *Dicer* in early neural development is not clear because *Dicer*-deficient mice die at embryonic day 7.5, before neurulation occurs (Murchison et al. 2005). However, conditional knockout of *Dicer* disrupts cellular and tissue morphogenesis in the cortex and hippocampus (HPC; Davis et al. 2008). Similarly, progressive cell death was observed when *Dicer* was inactivated postnatally in the cerebellum (Schaefer et al. 2007) or in dopaminergic neurons in the forebrain (Kim et al. 2007). These findings are consistent with a general role for miRNAs in cell survival and differentiation in the CNS. However, because *Dicer* is involved in both miRNA biogenesis and the siRNA machinery, a causal link between miRNAs and the observed phenotype in mammalian *Dicer* knockout animals has not been formally demonstrated.

Another important control of miRNA biogenesis is *Drosha* and its partner *Dgcr8*. *Dgcr8* homozygous knockout mice die at embryonic day 6.5 (Wang et al. 2007; Stark et al. 2008). *Dgcr8* heterozygous mice ($Dgcr8^{+/-}$) display partially impaired miRNA biogenesis (Stark et al. 2008; see below). Many, but not all, pri-miRNAs accumulate, whereas a subset of mature miRNAs are down-regulated due to haploinsuffiency of Dgcr8 levels in the HPC and prefrontal cortex (PFC) of mutant mice. Further investigation of HPC neurons indicated that the $Dgcr8^{+/-}$ mice had smaller dendritic spines and reduced dendritic complexity in the CA1 neurons. In addition, behavioral analysis showed deficits in cognitive performance and sensorimotor gating (Stark et al. 2008; see below).

The third checkpoint of miRNA maturation is the RISC pathway. The loading of the miRNA into the RISC is facilitated by Argonaute proteins (Agos). With a loss-of-function mutation of the Drosophila homologue of *AGO1*, *dAgo1*, the maternal and zygotic mutant embryos showed global developmental defects, with the most prominent malformation occurring in the nervous system (Kataoka et al. 2001). It has also been reported that *Ago2*-null mice die early in development and show severe neural tube defects (Liu et al. 2004). Interestingly, a more recent study shows that all Argonautes elevate mature miRNA expression post-transcriptionally, independent of RNase activity, suggesting coordinated regulation of miRNA expression and function (Diederichs and Haber 2007). Another important component of the RISC complex is the fragile X mental retardation protein (FMRP; Jin et al. 2004). FMRP is a RNA-binding protein that associates with polyribosomes. The hypermethylation of its CpG island within the coding region leads to inactivation of the gene in fragile X patients with mental retardation.

4.2 The impact of disrupting individual miRNAs

4.2.1 MiR-134 cluster on dendritic morphology and synaptic plasticity

The activity-dependent local protein synthesis in dendrites of neurons is important for synaptic plasticity as well as learning and memory function in the HPC (Sutton and Schuman 2006). The possibility that miRNAs may contribute to synapse formation and synaptic function by regulating the local translation of their target mRNAs is an area of active investigation. Indeed, miR-134 has been demonstrated to localize near synaptic sites in dendrites of HPC neurons and regulates the size of dendritic spines (Schratt et al. 2006). miR-134 can negatively regulate the width of dendritic spines but not their density, and it has no effect on dendritic branching. Overexpression of miR-134 significantly decreased the spine volume whereas 2′-*O*-methylated anti-miR-134 oligonucleotide increased spine width by 7.6%. Considering the heterogeneity and dynamic nature of dendritic spines on cultured neurons, this phenotype is relatively subtle and consistent with a modulatory role of miR-134 in spine formation. In searching for the potential target of miR-134, Schratt et al. (2006) identified Limk1, which regulates actin and microtubule polymerization, as a main target. Moreover, the negative effect of miR-134 overexpression on spine volume could be partially rescued by overexpression of Limk1. BDNF promotes dendritic spine growth and regulates synaptic function. The authors showed that BDNF treatment significantly increased synthesis of Limk1 protein. Interestingly, BDNF treatment relieved miR-134-dependent inhibition of Limk1 translation, whereas introduced synthetic miR-134 into neurons partially interfered with BDNF induction of Limk1 translation (Schratt et al. 2006). A further investigation into the transcription of a large cluster of brain-specific miRNAs, including miR-134, indicated that the miRNA gene cluster expression can be induced by increasing neuronal activity in primary rat neurons. The TF myocyte enhancing factor 2 (Mef2) is necessary and sufficient for controlling the activity-dependent expression of this miRNA cluster. Mef2 induces expression of miR-134 and promotes neurite outgrowth by inhibiting translation of the mRNA encoding for the translational repressor Pumilio2 (Fiore et al. 2009).

4.2.2 MiR-219 and NMDA receptor-mediated neurobehavioral dysfunction

A recent study showed that pharmacological (via dizocilpine) or genetic (*NR1* hypomorphism) disruption of NMDA receptor signaling leads to the decrease of miR-219 in the PFC of mice. In vivo inhibition of miR-219 by specific anti-miR in the murine brain caused up-regulation of its target calcium/calmodulin-dependent protein kinase II gamma subunit (CaMKII gamma). The abnormal expression of CAMKII gamma resulted in malfunction of NMDA receptor signaling and altered the corresponding behavioral responses. Interestingly, the dizocilpine-induced effects on miR-219 could be prevented by pretreating the mice with the antipsychotic drugs haloperidol and clozapine (Kocerha et al. 2009).

5 miRNA Dysregulation in Neuropsychiatric Disorders

Alterations in neuronal morphology and function underlie many forms of neurological and psychiatric disorders. Given that miRNAs regulate neuronal morphology and functions at multiple levels, it can be hypothesized that aberrant miRNA function might contribute to many of these disorders.

5.1 Schizophrenia

Several lines of evidence have emerged that miRNAs might be involved in the etiology of schizophrenia (SCZ). The strongest line of evidence for a link between

SCZ and miRNA biogenesis is provided by a recent study from our group using a mouse model of the 22q11.2 microdeletion (Stark et al. 2008).

5.1.1 22q11.2 microdeletion and risk for SCZ

Hemizygous microdeletions of the 22q11.2 locus are among the most common chromosomal abnormalities (1:4,000 live births; Botto et al. 2003). 22q11.2 microdeletions account for up to 1–2% of non-familial (sporadic) SCZ cases and represent the only confirmed recurrent copy number mutation responsible for introducing new cases of SCZ in the population (Karayiorgou et al. 1995; Xu et al. 2008; International SCZ Consortium 2008; Stefansson et al. 2008). It has been shown conclusively that ~30% of children with the 22q11.2 microdeletion will develop SCZ or schizoaffective disorder in adolescence or early adulthood (Pulver et al. 1994; Murphy et al. 1999). Importantly, a series of studies has revealed that SCZ patients carrying the 22q11.2 deletion bear the hallmark neuropsychological and neuroanatomical features of classical SCZ (Chow et al. 1999, 2002).

Individuals with the 22q11.2 microdeletion exhibit a spectrum of cognitive deficits in working memory, executive visual attention, visuospatial short-term memory, conflict monitoring, and numerical cognitive impairment (Bearden et al. 2001; Sobin et al. 2005; Lewandowski et al. 2007) and have decreased activation of the dorsolateral (dl) PFC during such tasks (Kates et al. 2007). This pattern of cognitive dysfunction is notable, especially given the increasing recognition of these cognitive impairments as contributing factors in SCZ-associated disability.

22q11.2 deletions are mediated by segmental duplications. The majority of them (87%) are 3-Mb in size, whereas a smaller percentage (8%) involves the same proximal breakpoint but a different distal breakpoint resulting in a smaller 1.5-Mb deletion (Edelmann et al. 1999; Fig. 2). Several schizophrenic patients have been described to carry the smaller 22q11.2 deletion (Karayiorgou et al. 1995; Xu et al. 2008; International SCZ Consortium 2008; Fig. 2). Most of the genes in this region are expressed in the brain in a relatively wide pattern. The results of human genetic analysis so far suggest that it is a rather small subset of the deleted genes and their interactions that contribute to the dramatic increase in disease risk (Liu et al. 2002a, b; Mukai et al. 2004; Paterlini et al. 2005; Mukai et al. 2008; Stark et al. 2008).

5.1.2 Generation of the 22q11.2 microdeletion mouse model(*Df(16)A*$^{+/-}$)

With the exception of *CLTCL1*, all of the functional human genes from the 22q11.2 locus shown in Figure 2 are represented in the mouse, although organized in a slightly different order. We used a Cre/Lox-dependent chromosomal engineering strategy to generate a mouse model carrying a 1.3-Mb chromosomal deletion, (*Df (16)A*$^{+/-}$), that spans a segment of the murine chromosome 16 syntenic to the previously defined 1.5-Mb critical region for SCZ (Stark et al. 2008). This deletion

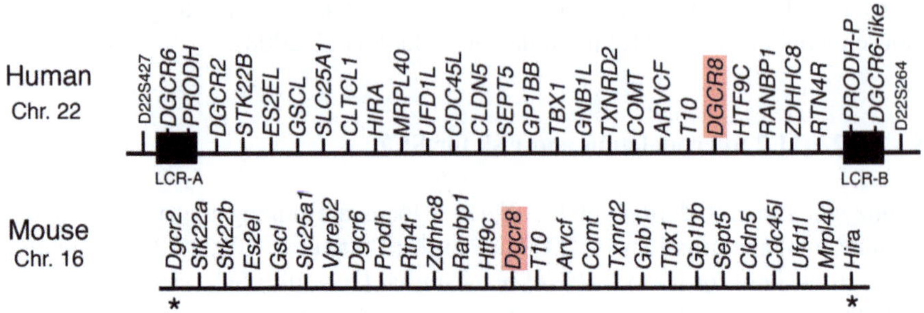

Fig. 2 The human 1.5-Mb 22q11.2 locus and the syntenic mouse region. Schematic diagram showing the 1.5-Mb human chromosome 22q11.2 region, located between microsatellite markers D22S427 and D22S264, and the syntenic mouse region. Almost all of the functional human genes in this segment are represented in the mouse, organized in a slightly different order. The 1.5-Mb deletion is mediated by low copy repeat sequences LCR-A and LCR-B (illustrated as black boxes). *PRODH-P* and *DGCR6-like* indicate pseudogenes. *Dgcr2* and *Hira*, the two endpoints of the targeted deletion, are indicated by asterisks

ranges from *Dgcr2* to *Hira* and encompasses 27 genes (Fig. 2). Mice harboring the deletion appeared normal from gross observation, although the number of *Df(16)A*$^{+/-}$ mice per litter at weaning age was ~30%, as opposed to the expected 50% transmission rate normally obtained from Heterozygote x WT crosses. Hemizygous *Df(16)A*$^{+/-}$ mice, however, did not show any gross brain abnormalities.

5.1.3 *Df(16)A*$^{+/-}$ mice: altered synaptic connectivity and cognitive function

Df(16)A$^{+/-}$ mice were hyperactive compared to WT littermates and more fearful of exploring a novel environment. In addition, *Df(16)A*$^{+/-}$ mice showed abnormalities in sensorimotor gating, as assayed by the pre-pulse inhibition (PPI) test, as well as deficits in associative and spatial learning and memory, as assayed by the contextual fear-conditioning and Morris Water-maze tests, respectively (Stark et al. 2008; unpublished data). Notably, spatial working memory-dependent learning was also affected in *Df(16)A*$^{+/-}$ mice (Stark et al. 2008). Although we did not find any gross anatomical abnormalities in the *Df(16)A*$^{+/-}$ mice, behavioral and cognitive deficits were accompanied by specific neuronal pathology in the HPC. Specifically, *Df(16)A*$^{+/-}$ mice showed a reduction in the mushroom spine density in CA1 pyramidal neurons compared to WT mice, whereas the density of other spine morphotypes was not affected. In addition, they showed a decrease in the spine width (but not length) of CA1 pyramidal neurons. The decrease in spine density was accompanied by a decrease in the density of glutamatergic synapses in vivo. Specifically, by using quantitative immunohistochemistry to measure the number of PSD95 and VGLUT1 puncta in both apical (SR: stratum radiatum) and basal (SO: stratum oriens) dendrites of pyramidal cells sampled along the dorsal/ventral axis of CA1, we found a significant reduction in the number of puncta in both layers in the *Df(16)A*$^{+/-}$ mice. Finally, *Df(16)A* affected dendritic complexity, resulting in a simplification of the dendritic complexity of CA1 pyramidal neurons. Notably, many of the deficits observed in the HPC of adult mice were recapitulated in primary cultures of embryonic hippocampal neurons.

5.1.4 Abnormal miRNA processing in *Df(16)A*$^{+/-}$ mice

To obtain a comprehensive view of the molecular pathways affected in vivo by the deficiency of the 22q11.2 orthologues, we resorted to an unbiased evaluation of the transcriptional programs affected in PFC and HPC of the *Df(16)A*$^{+/-}$ mice. We focused on PFC and HPC because of a diverse and convergent body of studies suggesting functional and structural pathology of these brain regions in individuals with the 22q11.2 deletions (Meechan et al. 2006; Sivagnanasundaram et al. 2007). In addition to the expected down-regulation of the genes within the deletion region, analysis of GeneChip data (Affymetrix) at stringent false discovery rate (FDR) P-value < 0.01 identified 837 probe sets (1.9%) in PFC and 96 probe sets (0.21%) in HPC as being differentially expressed. Interestingly, there was only partial overlap in the transcriptional response between the two brain regions, consisting of 18 probe sets (FDR P-value < 0.01) located outside the deficiency. Intriguingly, 6 of the 18 common probe sets (33%, all up-regulated) were located in the immediate vicinity of genes containing miRNAs, presumably within precursor pri-miRNA forms. The observation that probe sets overlapping with known miRNA locations were among the top-scoring ones in our expression profiling analysis highlighted the relative importance of these changes in the context of the entire transcriptome.

We speculated that the consistent up-regulation of miRNA-related transcripts that we observed reflected, at least in part, the up-regulation of the pri-forms of miRNAs, due to the fact that one of the disrupted 22q11.2 genes is *Dgcr8*, an important component of the "microprocessor" complex (see above). Indeed, we provided compelling evidence for an abnormality in brain miRNA processing in *Df (16)A*$^{+/-}$ mice, emerging as a result of the down-regulation of the *Dgcr8* gene (Stark et al. 2008). The end effect of *Dgcr8* haploinsufficiency is the down-regulation of a specific subset of mature miRNAs that we identified by miRNA profiling analysis. Why all miRNAs were not affected is unclear. In addition to miRNA production via non-canonical pathways (see above), this may reflect differences in enzymatic processing among individual pri-miRNAs or other yet unidentified compensations. Interestingly, among the affected miRNAs is miR-134 (see above). It should be noted that, in the majority of cases, the expression of miRNAs is reduced by ~20–70%. Recent studies strongly suggest that partial reduction in miRNA levels within this range can have an effect on transcript or protein levels of target genes. In addition, since target mRNAs can be simultaneously bound and repressed by more than one miRNA species, in a manner that allows for cooperativity between target sites, the level of repression achieved may be highly sensitive to the amount of available miRNA complexes and the number of affected miRNAs (Hobert 2008; Stark et al. 2008). Consistent with the expectation that synchronous modest decreases in a number of miRNAs may have considerable impact on target transcript stability, we estimated that impaired miRNA biogenesis accounts for at least a portion of the transcript up-regulation observed in the PFC and HPC of the *Df(16)A*$^{+/-}$ mice.

Specifically, we identified 59 miRNA genes ($\sim$19%) in PFC and 30 ($\sim$10%) in HPC that showed significant down-regulation (ratio *Df(16)A*$^{+/-}$ / WT < 0.8, FDR P-value < 0.05; Stark et al. 2008). Notably, among them, 25 miRNAs were shared between PFC and HPC. We used TargetScan (version 4) and PicTar to identify putative seed sites for all down-regulated mature miRNAs within the 3′-UTR of all unique known GeneChip transcripts within the TargetScan or PicTar databases, which are differentially expressed in the PFC or HPC. miRNAs can modulate transcript stability and their down-regulation is expected to result in up-regulation of target genes. Consistent with this expectation, we found a higher proportion of up-regulated PFC and HPC transcripts to contain one or more potential seed sites for any of the affected miRNAs, as compared to down-regulated transcripts. A frequency distribution analysis showed that this enrichment was exclusively confined in up-regulated transcripts containing multiple (≥ 2) seed sites (10–20% of the up-regulated transcripts in either PFC or HPC), indicating possible convergent /cooperative effects of the affected miRNAs in regulating the stability of individual transcripts.

5.1.5 Dgcr8 deficiency affects neuronal development and cognitive function

To examine the functional consequences of alterations in miRNA biogenesis in *Df (16)A*$^{+/-}$ mice (see above) and identify the miRNA regulated genes, we generated *Dgcr8*-deficient mice using an embryonic stem cell line carrying a b-geo gene trap inserted into intron 8 of the *Dgcr8* gene (see Stark et al. 2008). Homozygous mice die before birth (unpublished data) but heterozygous mice (*Dgcr8*$^{+/-}$) are produced at the expected rate, appear physically healthy and show normal gross brain morphology.

We showed that haploinsufficiency at the *Dgcr8* locus and the ensuing alterations in miRNA biogenesis contribute to the behavioral and cognitive deficits observed in the *Df(16)A*$^{+/-}$ mice. *Dgcr8*-deficient mice had impaired acquisition of a spatial working memory-dependent task and impaired sensorimotor gating. By contrast, they had normal associative memory, as assayed by the contextual and cued fear conditioning test. These findings demonstrate for the first time that abnormal miRNA biogenesis affects cognitive performance in mice. Notably, in *Dgcr8*$^{+/-}$ mice only spatial working memory-based learning appears to be affected, whereas in *Df(16)A*$^{+/-}$ mice associative learning is also compromised. The difference between the two cognitive tasks could be related to disparities in the underlying cognitive demands, the complexity of the relevant neural circuits and their differential sensitivity to disruptions of miRNA biogenesis.

As mentioned above, miRNAs may contribute to synaptic development and maturation, providing, at least in part, a potential explanation for the cognitive and behavioral deficits observed in the *Dgcr8*$^{+/-}$ mice. Indeed, *Dgcr8*-deficient mice demonstrated impaired dendritic tree and dendritic spine development. Specifically, morphotypic analysis of mushroom spines showed that there was a significant decrease in the width of CA1 pyramidal spines, whereas their length was unaffected, a pattern very similar to the one we observed in the *Df(16)A*$^{+/-}$ mice. However, unlike in the *Df(16)A*$^{+/-}$ mice, we did not observe any changes in the mushroom spine density in the basal dendrites of CA1 pyramidal neurons from *Dgcr8*$^{+/-}$ mice compared to wild type mice. Finally, Sholl analysis indicated an effect of *Dgcr8* deficiency on dendritic complexity at varying distances from the soma, primarily at the distal dendrites.

5.1.6 Dgcr8 haploinsufficiency interacts with haploinsufficiency of other 22q11.2 genes

Complementary effects of impaired miRNA biogenesis (due to reduced dosage of *Dgcr8*) and impaired neuronal protein palmitoylation (due to reduced dosage of a neighboring gene, *Zdhhc8*, a member of the DHHC family of palmitoyltransferases) appear to underlie the impaired dendritic growth and spine development observed in the *Df(16)A*$^{+/-}$ mice. The complementary contribution of *Dgcr8 and Zdhhc8* haploinsufficiency to spine development is particularly informative in that respect. Notably, analysis of spine morphogenesis in vivo in the CA1 subfield of the HPC of *Df(16)A*$^{+/-}$ mice revealed a reduction in mushroom spine density (~25%) and width (~10%), but not length. Decreases in density (but not size) of the dendritic spines could be accounted for by the deficiency of the *Zdhhc8* gene (Mukai et al. 2008). In a complementary fashion, decreases in size (but not density) of the dendritic spines could be accounted for by the deficiency of the *Dgcr8* gene. Thus, *Dgcr8* haploinsufficiency, although not solely responsible, very likely interacts with haploinsufficiency of other 22q11.2 genes to produce the synaptic alterations, as well as the cognitive and behavioral deficits, associated with the 22q11.2 microdeletions. These interactions likely underlie the large increase in schizophrenia risk associated with the 22q11.2 microdeletions.

5.1.7 miRNAs and SCZ

Based on these findings, it is tempting to speculate that miRNA-dependent regulation of neuronal connectivity might contribute to the pathogenesis of SCZ and other psychiatric disorders. In addition, our findings could have more general implications for understanding the genetic basis of psychiatric disorders. In particular, our results suggest that the potential of miRNAs to contribute to the regulation of expression of multiple genes in the brain could be an important component of the genetically complex architecture of such disorders. Several recent lines of evidence appear to support these hypotheses.

Perkins et al. (2007) compared the expression profile of 264 human miRNAs from postmortem PFC of individuals with SCZ with that of unaffected individuals. They found 16 miRNAs to be differentially expressed: 15 showed decreased expression and one showed higher expression in SCZ patients as compared with controls at 5% FDR. They further compared expression levels of the miRNAs residing in introns of protein-coding genes and those of the host genes and observed that the ratios of microarray expression levels of four of the six such miRNAs were significantly different from those of the host genes in SCZ patients (Perkins et al. 2007). These results suggest that the alteration of miRNAs they observed is due to the change of miRNA biogenesis procedure rather than transcription of pri-miRNA. More recently, another expression profiling analysis of postmortem cortical gray matter from the superior temporal gyrus revealed significant up-regulation of miR-181b expression in SCZ. The result was confirmed by quantitative real-time polymerase chain reaction (qRT-PCR) analysis of miRNA expression in a cohort of 21 matched pairs of SCZ and non-psychiatric controls. This study provided supporting evidence that altered miRNA levels could be a significant factor in the dysregulation of cortical gene expression in SCZ (Beveridge et al. 2008).

In addition, a SNP-based genetic analysis identified an association of two miRNA loci, miR-206 and miR-198, with SCZ in the Danish and Norwegian sample (Hansen et al. 2007), whereas a genome-wide scan for copy number mutations in sporadic SCZ identified a de novo duplication encompassing the *DICER1* gene (Xu et al. 2008). Although further validation of these correlations is necessary, these studies further support a connection between miRNAs and the genetic architecture of SCZ.

5.2 Tourette's syndrome

Tourette's syndrome (TS) is a developmental neuropsychiatric disorder characterized by chronic vocal and motor tics. The first report linking miRNAs to TS came from a genetic association study and mutational screening of a rare subset of TS patients with chromosomal anomalies on a TS candidate gene, Slit and Trk-like1 (*SLITRK1*), that encodes for a protein important for growth, guidance and branching of neuronal processes. Genetic association analysis revealed a correlation of mutations within the miR-189 binding site in the 3'UTR of the *SLITRK1* gene and increased disease risk. Further studies indicated that the mutation leads to the replacement of one GU wobble base pairing with an AU base pairing, resulting in a stronger interaction of the miRNA with the target mRNA. Since TS is a complex disease, the polymorphism identified probably represents only one of several mutations that contribute to the phenotype. Nevertheless, this mutation demonstrated the possible relevance of the interaction between miRNA and target mRNA under the physiological and pathological conditions (Abelson et al. 2005).

5.3 Autism spectrum disorders

Initial evidence has also been provided that miRNAs might play a role in autism spectrum disorders (ASDs). By using multiplex q-PCR, Abu-Elneel et al. (2008) compared the expression of 466 human miRNAs from 13 postmortem cerebellar cortex tissues from individuals with ASDs and 13 non-autistic controls. While most miRNA levels failed to show any significant difference across all samples, several miRNAs, such as miR-320 and miR-598, were differentially expressed in the autistic samples compared to the mean control value, suggesting that alterations of some miRNAs may contribute to the autism phenotypic variation (Abu-Elneel et al. 2008). In another study, by using lymphoblastoid cell line samples, Talebizadeh et al. (2008) checked the expression profile of 470 mature human miRNAs of six ASD patients and six matched control. Nine of the 470 miRNA genes showed differential expression (either higher or lower) in the ASD samples compared with controls. These results suggest a potential contribution of miRNAs to ASDs (Talebizadeh et al. 2008). This finding is further supported by recent findings in Rett syndrome (RTT), an X-linked neurodevelopmental disorder occurring primarily in females and characterized by developmental stagnation, stereotypical movements, microcephaly, seizures and autistic features. RTT is caused by mutations in the gene encoding methyl-CpG binding protein 2 (MeCP2; Amir et al. 1999). Recently, a brain-enriched miRNA, miR-132, has been shown to regulate the expression of MeCP2 (Klein et al. 2007). It was shown that miR-132 is both necessary and sufficient to regulate MeCP2 protein levels in neurons. This result, combined with previous results that BDNF induced miR-132 transcription and lack of MeCP2 has decreased BDNF levels in mouse models of RTT, suggests that miR-132 might act as a homeostatic controller of MeCP2 translation. The abnormalities of miR-132 lead to a deregulation of MeCP2 and contribute to RTT. Another study reported that expression of miR-184, a brain-specific miRNA, is repressed by the binding of MeCP2 to its promoter. The depolarization released the repression of MeCP2 and caused paternal allele-specific expression of miR-184. This finding provided support for the link between the miRNA and DNA methylation pathways and RTT (Nomura et al. 2008).

References

Abelson JF, Kwan KY, O'Roak BJ, Baek DY, Stillman AA, Morgan TM, Mathews CA, Pauls DL, Rasin MR, Gunel M, Davis NR, Ercan-Sencicek AG, Guez DH, Spertus JA, Leckman JF, Dure LS 4th, Kurlan R, Singer HS, Gilbert DL, Farhi A, Louvi A, Lifton RP, Sestan N, State MW (2005) Sequence variants in SLITRK1 are associated with Tourette's syndrome. Science 310:317–320

Abu-Elneel K, Liu T, Gazzaniga FS, Nishimura Y, Wall DP, Geschwind DH, Lao K, Kosik KS (2008) Heterogeneous dysregulation of microRNAs across the autism spectrum. Neurogenetics 9:153–161

Amir RE, Van den Veyver IB, Wan M, Tran CQ, Francke U, Zoghbi HY (1999) Rett syndrome is caused by mutations in X-linked MECP2, encoding methyl-CpG-binding protein 2. Nature Genet 23:185–188

Babiarz JE, Ruby JG, Wang Y, Bartel DP, Blelloch R (2008) Mouse ES cells express endogenous shRNAs, siRNAs, and other Microprocessor-independent, Dicer-dependent small RNAs Genes Dev 22:773–2785

Baek D, Villen J, Shin C, Camargo FD, Gygi SP, Bartel DP (2008) The impact of microRNAs on protein output. Nature 455:64-71

Bartel DP, Chen CZ (2004) Micromanagers of gene expression: the potentially widespread influence of metazoan microRNAs. Nature Rev Genet 5:396–400

Bearden CE, Woodin MF, Wang PP, Moss E, McDonald-McGinn D, Zackai E, Emannuel B, Cannon TD (2001) The neurocognitive phenotype of the 22q112 deletion syndrome: selective deficit in visual-spatial memory. J Clin ExpNeuropsychol 23:447–464

Behm-Ansmant I, Rehwinkel J, Doerks T, Stark A, Bork P, Izaurralde E (2006) mRNA degradation by miRNAs and GW182 requires both CCR4:NOT deadenylase and DCP1:DCP2 decapping complexes. Genes Dev 20:1885–1898

Beveridge NJ, Tooney PA, Carroll AP, Gardiner E, Bowden N, Scott RJ, Tran N, Dedova I, Cairns MJ (2008) Dysregulation of miRNA 181b in the temporal cortex in schizophrenia. Human Mol Genet 17:1156–1168

Borchert, GM, Lanier W, Davidson BL (2006) RNA polymerase III transcribes human microRNAs. Nature Struct Mol Biol 13:1097–1101

Botto LD, May K, Fernhoff PM, Correa A, Coleman K, Rasmussen SA, Merritt RK, O'Leary LA, Wong LY, Elixson EM, Mahle WT, Campbell RM (2003) A population-based study of the 22q112 deletion: phenotype, incidence, and contribution to major birth defects in the population. Pediatrics 112:101–107

Chen JF, Mandel EM, Thomson JM, Wu Q, Callis TE, Hammond SM, Conlon FL, Wang DZ (2006) The role of microRNA-1 and microRNA-133 in skeletal muscle proliferation and differentiation. Nature Genet 38:228–233

Chow EW, Mikulis DJ, Zipursky RB, Scutt LE, Weksberg R, Bassett AS (1999) Qualitative MRI findings in adults with 22q11 deletion syndrome and schizophrenia. Biol Psychiat 46:1436–1442

Chow EW, Zipursky RB, Mikulis DJ, Bassett AS (2002) Structural brain abnormalities in patients with schizophrenia and 22q11 deletion syndrome. Biol Psychiat 51:208–215

Cohen SM, Brennecke J, Stark A (2006) Denoising feedback loops by thresholding–a new role for microRNAs. Genes Dev 20:2769–2772

Davis TH, Cuellar TL, Koch SM, Barker AJ, Harfe BD, McManus MT, Ullian EM (2008) Conditional loss of Dicer disrupts cellular and tissue morphogenesis in the cortex and hippocampus. J Neurosci 28:4322–4330

Diederichs S, Haber DA (2007) Dual role for argonautes in microRNA processing and posttranscriptional regulation of microRNA expression. Cell 131:1097–1108

Edelmann L, Pandita RE, Morrow BE (1999) Low-copy repeats mediate the common 3-Mb deletion in patients with velo-cardio-facial syndrome. Am J Human Genet 64:1076–1086

Eulalio A, Behm-Ansmant I, Schweizer D, Izaurralde E (2007) P-body formation is a consequence, not the cause, of RNA-mediated gene silencing. Mol Cell Biol 27:3970–3981

Fazi F, Rosa A, Fatica A, Gelmetti V, De Marchis ML, Nervi C, Bozzoni I (2005) A minicircuitry comprised of microRNA-223 and transcription factors NFI-A and C/EBPalpha regulates human granulopoiesis. Cell 123:819–831

Fiore R, Khudayberdiev S, Christensen M, Siegel G, Flavell SW, Kim TK, Greenberg ME, Schratt G (2009) Mef2-mediated transcription of the miR379-410 cluster regulates activity-dependent dendritogenesis by fine-tuning Pumilio2 protein levels. EMBO J 28:697–710

Giraldez AJ, Cinalli RM, Glasner ME, Enright AJ, Thomson JM, Baskerville S, Hammond SM, Bartel DP, Schier AF (2005) MicroRNAs regulate brain morphogenesis in zebrafish. Science 308:833–838

Hansen T, Olsen L, Lindow M, Jakobsen KD, Ullum H, Jonsson E, Andreassen OA, Djurovic S, Melle I, Agartz I, Hall H, Timm S, Wang AG, Werge T (2007) Brain expressed microRNAs implicated in schizophrenia etiology. PLoS ONE 2:e873

Hobert O (2008) Gene regulation by transcription factors and microRNAs. Science 319:1785–1786

Humphreys DT, Westman BJ, Martin DI, Preiss T (2005) MicroRNAs control translation initiation by inhibiting eukaryotic initiation factor 4E/cap and poly(A) tail function. Proc Natl Acad Sci USA 102:16961–16966

Hutvágner G, McLachlan J, Pasquinelli AE, Bálint E, Tuschl T, Zamore PD (2001) A cellular function for the RNA-interference enzyme Dicer in the maturation of the let-7 small temporal RNA. Science 293:834–838

International Schizophrenia Consortium (2008) Rare chromosomal deletions and duplications increase risk of schizophrenia. Nature 455:237–241

Jin P, Zarnescu DC, Ceman S, Nakamoto M, Mowrey J, Jongens TA, Nelson DL, Moses K, Warren ST (2004) Biochemical and genetic interaction between the fragile X mental retardation protein and the microRNA pathway. Nature Neurosci 7:113–117

Karayiorgou M, Morris MA, Morrow B, Shprintzen RJ, Goldberg R, Borrow J, Gos A, Nestadt G, Wolyniec PS, Lasseter VK (1995) Schizophrenia susceptibility associated with interstitial deletions of chromosome 22q11. Proc Natl Acad Sci USA 92:7612–7616

Kataoka Y, Takeichi M, Uemura T (2001) Developmental roles and molecular characterization of a Drosophila homologue of Arabidopsis Argonaute1, the founder of a novel gene superfamily. Genes Cells 6:313–325

Kates WR, Antshel KM, Fremont WP, Shprintzen RJ, Strunge LA, Burnette CP, Higgins AM (2007) Comparing phenotypes in patients with idiopathic autism to patients with velocardiofacial syndrome (22q11 DS) with and without autism. Am J Med Genet Part A 143A:2642–2650

Kim VN (2005) MicroRNA biogenesis: coordinated cropping and dicing. Nature Rev Mol Cell Biol 6:376–385

Kim J, Inoue K, Ishii J, Vanti WB, Voronov SV, Murchison E, Hannon G, Abeliovich A (2007) A microRNA feedback circuit in midbrain dopamine neurons. Science 317:1220-1224

Klein ME, Lioy DT, Ma L, Impey S, Mandel G, Goodman RH (2007) Homeostatic regulation of MeCP2 expression by a CREB-induced microRNA. Nature Neurosci 10:1513–1514

Kocerha J, Faghihi MA, Lopez-Toledano MA, Huang J, Ramsey AJ, Caron MG, Sales N, Willoughby D, Elmen J, Hansen HF, Orum H, Kauppinen S, Kenny PJ, Wahlestedt C (2009) MicroRNA-219 modulates NMDA receptor-mediated neurobehavioral dysfunction. Proc Natl Acad Sci USA 106:3507–3512

Krichevsky AM, Sonntag KC, Isacson O, Kosik KS (2006) Specific microRNAs modulate embryonic stem cell-derived neurogenesis. Stem Cells 24:857–864

Lewandowski KE, Shashi V, Berry PM, Kwapil TR (2007) Schizophrenic-like neurocognitive deficits in children and adolescents with 22q11 deletion syndrome. Am J Med Genet B Neuropsychiatr Genet 144B:27–36

Liu H, Heath SC, Sobin C, Roos JL, Galke BL, Blundell ML, Lenane M, Robertson B, Wijsman EM, Rapoport JL, Gogos JA, Karayiorgou M (2002a) Genetic variation at the 22q11 PRODH2/DGCR6 locus presents an unusual pattern and increases susceptibility to schizophrenia. Proc Natl Acad Sci USA 99:3717–3722

Liu H, Abecasis GR, Heath SC, Knowles A, Demars S, Chen YJ, Roos JL, Rapoport JL, Gogos JA, Karayiorgou M (2002b) Genetic variation in the 22q11 locus and susceptibility to schizophrenia. Proc Natl Acad Sci USA 99:16859–16864

Liu J, Carmell MA, Rivas FV, Marsden CG, Thomson JM, Song JJ, Hammond SM, Joshua-Tor L, Hannon GJ (2004) Argonaute2 is the catalytic engine of mammalian RNAi. Science 305:1437–1441

Makeyev E, Maniatis T (2008) Multilevel regulation of gene expression by microRNAs. Science 319:1789–1790

Maroney PA, Yu Y, Fisher J, Nilsen TW (2006) Evidence that microRNAs are associated with translating messenger RNAs in human cells. Nature Struct Mol Biol 13:1102–1107

Meechan DW, Maynard TM, Wu Y, Gopalakrishna D, Lieberman JA, LaMantia AS (2006) Gene dosage in the developing and adult brain in a mouse model of 22q11 deletion syndrome. Mol Cell Neurosci 33:412–428

Mukai J, Liu H, Burt RA, Swor DE, Lai WS, Karayiorgou M, Gogos JA (2004) Evidence that the gene encoding ZDHHC8 contributes to the risk of schizophrenia. Nature Genet 36:725–731

Mukai J, Dhilla A, Drew LJ, Stark KL, Cao L, MacDermott AB, Karayiorgou M, Gogos JA (2008) Palmitoylation-dependent neurodevelopmental deficits in a mouse model of 22q11 microdeletion. Nature Neurosci 11:1302–1310

Murchison EP, Partridge JF, Tam OH, Cheloufi S, Hannon GJ (2005) Characterization of Dicer-deficient murine embryonic stem cells. Proc Natl Acad Sci USA 102: 12135–12140

Murphy KC, Jones LA, Owen MJ (1999) High rates of schizophrenia in adults with velo-cardio-facial syndrome. Arch Gen Psychiat 56:940–945

Nomura T, Kimura M, Horii T, Morita S, Soejima H, Kudo S, Hatada I (2008) MeCP2-dependent repression of an imprinted miR-184 released by depolarization. Human Mol Genet 17:1192–1199

Obernosterer G, Leuschner PJ, Alenius M, Martinez J (2006) Post-transcriptional regulation of microRNA expression. RNA 12:1161–1167

O'Donnell KA, Wentzel EA, Zeller KI, Dang CV, Mendell JT (2005) c-Myc-regulated microRNAs modulate E2F1 expression. Nature 435:839–843

Okamura K, Hagen JW, Duan H, Tyler DM, Lai EC (2007) The mirtron pathway generates microRNA-class regulatory RNAs in Drosophila. Cell 130:89–100

Paterlini M, Zakharenko SS, Lai WS, Qin J, Zhang H, Mukai J, Westphal KG, Olivier B, Sulzer D, Pavlidis P, Siegelbaum SA, Karayiorgou M, Gogos JA (2005) Transcriptional and behavioral interaction between 22q112 orthologs modulates schizophrenia-related phenotypes in mice. Nature Neurosci 8:1586–1594

Perkins DO, Jeffries CD, Jarskog LF, Thomson JM, Woods K, Newman MA, Parker JS, Jin J, Hammond SM (2007) microRNA expression in the prefrontal cortex of individuals with schizophrenia and schizoaffective disorder. Genome Biol 8:R27

Petersen CP, Bordeleau ME, Pelletier J, Sharp PA (2006) Short RNAs repress translation after initiation in mammalian cells. Mol Cell 21:533–542

Pillai RS, Artus CG, Filipowicz W (2004) Tethering of human Ago proteins to mRNA mimics the miRNA-mediated repression of protein synthesis. RNA 10:1518–1525

Pulver AE, Nestadt G, Goldberg R, Shprintzen RJ, Lamacz M, Wolyniec PS, Morrow B, Karayiorgou M, Antonarakis SE, Housman D, Kucheriapati R (1994) Psychotic illness in patients diagnosed with velo-cardio-facial syndrome and their relatives. J Nerv Ment Dis 182:476–478

Ruby JG, Jan CH, Bartel DP (2007) Intronic microRNA precursors that bypass Drosha processing. Nature 448:83–86

Schaefer A, O'Carroll D, Tan CL, Hillman D, Sugimori M, Llinas R, Greengard P (2007) Cerebellar neurodegeneration in the absence of microRNAs. J Exp Med 204:1553–1558

Schratt GM, Tuebing F, Nigh EA, Kane CG, Sabatini ME, Kiebler M, Greenberg ME (2006) A brain-specific microRNA regulates dendritic spine development. Nature 439:283–289

Schulman BR, Esquela-Kerscher A, Slack FJ (2005) Reciprocal expression of lin-41 and the microRNAs let-7 and mir-125 during mouse embryogenesis. Dev Dyn 234:1046–1054

Selbach M, Schwanhausser B, Thierfelder N, Fang Z, Khanin R, Rajewsky N (2008) Widespread changes in protein synthesis induced by microRNAs. Nature 455:58–63

Sivagnanasundaram S, Fletcher D, Hubank M, Illingworth E, Skuse D, Scambler P (2007) Differential gene expression in the hippocampus of the Df1/+ mice: a model for 22q112 deletion syndrome and schizophrenia. Brain Res 1139:48–59

Sobin C, Kiley-Brabeck K, Karayiorgou M (2005) Lower prepulse inhibition in children with the 22q11 deletion syndrome. Am J Psychiatr 162:1090–1099

Stark KL, Xu B, Bagchi A, Lai WS, Liu H, Hsu R, Wan X, Pavlidis P, Mills AA, Karayiorgou M, Gogos JA (2008) Altered brain microRNA biogenesis contributes to phenotypic deficits in a 22q11-deletion mouse model. Nature Genet 40:751–760

Stefansson H, Rujescu D, Cichon S, Pietiläinen OP, Ingason A, Steinberg S, Fossdal R, Sigurdsson E, Sigmundsson T, Buizer-Voskamp JE, Hansen T, Jakobsen KD, Muglia P, Francks C, Matthews PM, Gylfason A, Halldorsson BV, Gudbjartsson D, Thorgeirsson TE, Sigurdsson A, Jonasdottir A, Jonasdottir A, Bjornsson A, Mattiasdottir S, Blondal T, Haraldsson M, Magnusdottir BB, Giegling I, Möller HJ, Hartmann A, Shianna KV, Ge D, Need AC, Crombie C, Fraser G, Walker N, Lonnqvist J, Suvisaari J, Tuulio-Henriksson A, Paunio T, Toulopoulou T, Bramon E, Di Forti M, Murray R, Ruggeri M, Vassos E, Tosato S, Walshe M, Li T, Vasilescu C, Mühleisen TW, Wang AG, Ullum H, Djurovic S, Melle I, Olesen J, Kiemeney LA, Franke B; GROUP, Sabatti C, Freimer NB, Gulcher JR, Thorsteinsdottir U, Kong A, Andreassen OA, Ophoff RA, Georgi A, Rietschel M, Werge T, Petursson H, Goldstein DB, Nöthen MM, Peltonen L, Collier DA, St Clair D, Stefansson K (2008) Large recurrent microdeletions associated with schizophrenia. Nature 455:232–236

Sutton MA, Schuman EM (2006) Dendritic protein synthesis, synaptic plasticity, and memory. Cell 127:49–58

Talebizadeh Z, Butler MG, Theodoro MF (2008) Feasibility and relevance of examining lymphoblastoid cell lines to study role of microRNAs in autism. Autism Res 1:307

Thomson JM, Newman M, Parker JS, Morin-Kensicki EM, Wright T, Hammond SM (2006) Extensive post-transcriptional regulation of microRNAs and its implications for cancer. Genes Dev 20:2202–2207

Vasudevan S, Tong Y, Steitz JA (2007) Switching from repression to activation: microRNAs can up-regulate translation. Science 318:1931–1934

Wang Y, Medvid R, Melton C, Jaenisch R, Blelloch R (2007) DGCR8 is essential for microRNA biogenesis and silencing of embryonic stem cell self-renewal. Nature Genet 39:380–385

Xu B, Roos JL, Levy S, van Rensburg EJ, Gogos JA, Karayiorgou M (2008) Strong association of de novo copy number mutations with sporadic schizophrenia. Nature Genet 40:880–885

Zhao Y, Samal E, Srivastava D (2005) Serum response factor regulates a muscle-specific microRNA that targets Hand2 during cardiogenesis Nature 436:214–220

Index

MIX
Papier aus verantwortungsvollen Quellen
Paper from responsible sources
FSC® C105338

If you have any concerns about our products,
you can contact us on
ProductSafety@springernature.com

In case Publisher is established outside the EU,
the EU authorized representative is:
Springer Nature Customer Service Center GmbH
Europaplatz 3, 69115 Heidelberg, Germany

Printed by Libri Plureos GmbH
in Hamburg, Germany